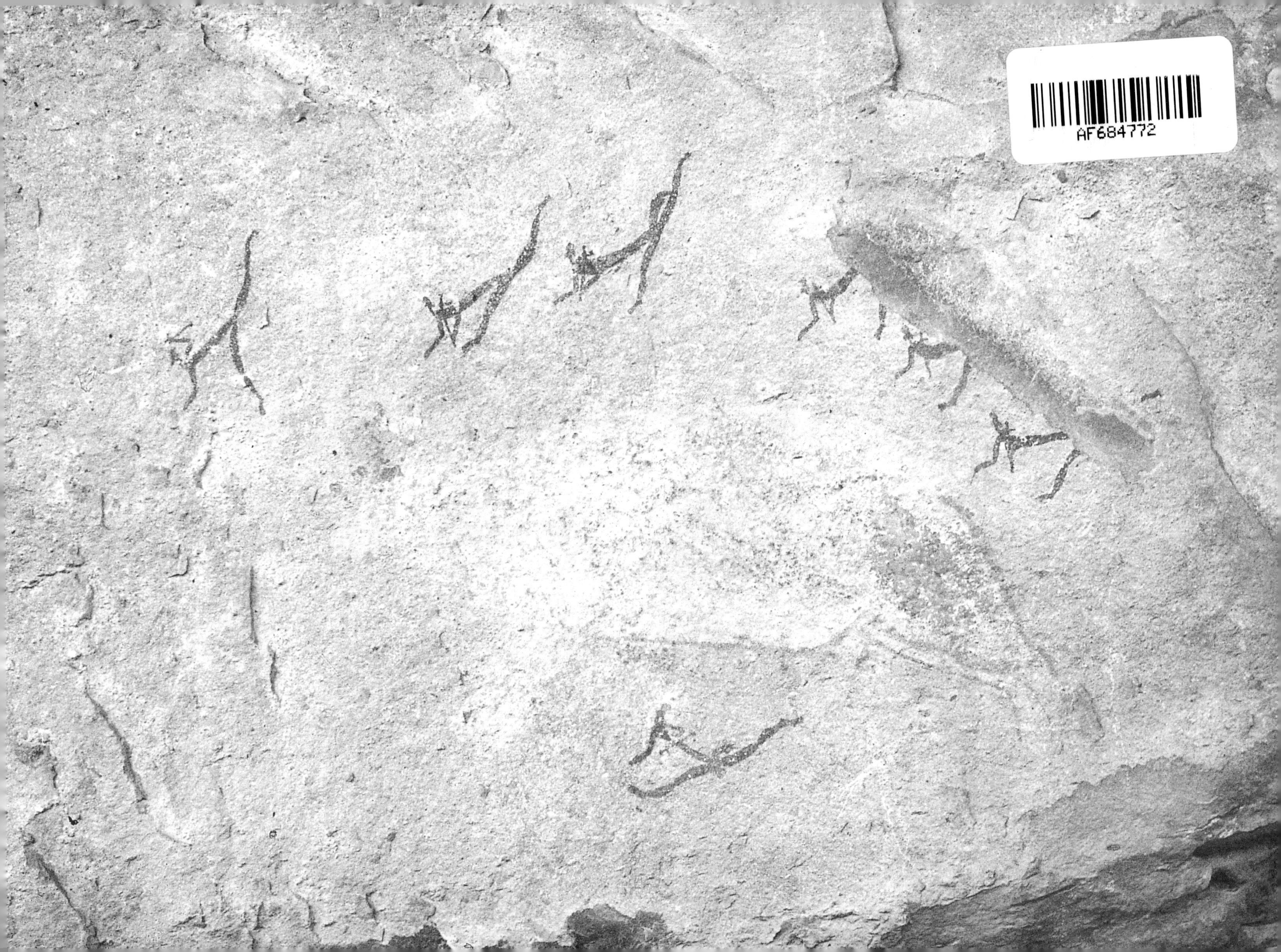

WW

Bert Woodhouse

THE RAIN AND ITS CREATURES

As the Bushmen painted them

WILLIAM WATERMAN PUBLICATIONS

For Amy

DIMENSIONS

The largest rain animals depicted in this book are those in the Wodehouse District (Figure 53) which are approximately one and a half metres long. Others vary between 200 and 600 mm (eight inches to two feet). The various rain snakes are approximately one metre long with accompanying human figures varying from 300 mm (one foot) down to 100 mm (four inches).

William Waterman Publications
A Division of Ashanti Publishing (Pty) Ltd
PO Box 5091
Rivonia 2128

ISBN 0 9583252 2 7

First Published 1992

Book design: Unifoto Cape Town
Typesetting and Reproduction: Unifoto
Printed and bound by Belmor Book Printers, Cape Town

Endpaper photograph: The rain animal ballet. Lesotho.
References page illustration: River system and source figure. Ndedema.

CONTENTS

> 'There is much beauty in the hills on the Buffalo River, with bush and cliffs, high mountains and ravines, where the only Bushmen known in Kaffirland dwelt in the "krantz", living in their early primitive condition, practising as rainmakers to the tribes, revered and wooed as prophets.'
>
> Major D B Hook, magistrate at Tsolo 1884-1886 in *With Sword and Statute* 1908.

AUTHOR'S FOREWORD

During the last seventy years or so many books have been published which have contributed to the familiarisation of the reading public with the art on the rocks of southern Africa. All have had their special merits and, although they have been profusely illustrated, there has been comparatively little overlap in the pictures. This is an indication of the wealth of interesting material available for publication.

Most of the books have dealt with the subject in general terms or on the basis of a particular geographical area. This book, like its predecessor *When Animals Were People* is an attempt to follow a particular theme — in this case rain and rainmaking — from one site to another throughout the sub-continent and, in so doing, to demonstrate the virtuosity of the individual artists when tackling the visual expression of a concept which was important to their culture.

This book concentrates on this theme in the hope that it will encourage readers to appreciate rock art generally and to find other themes of special interest which will enrich their travels both at home and abroad.

Because of the diversity of cultures in, and the variety of interpretations applied to, rock art, I have used some terminology in a specific sense. Instead of the word 'prophets' other writers have used such phrases as 'rain sorcerers', 'rain doctors', 'medicine men of the rain', or more recently, 'shamans'. In my narrative I have used all these descriptions at one time or another to suit the situation or the quotation. When dealing with the subject matter of a large proportion of the art concerned with rainmaking, I regard them as interchangeable.

The word 'shaman' was first used to denote a medicine man of the early religions of north-east Asia and of the American Indians, but it is now

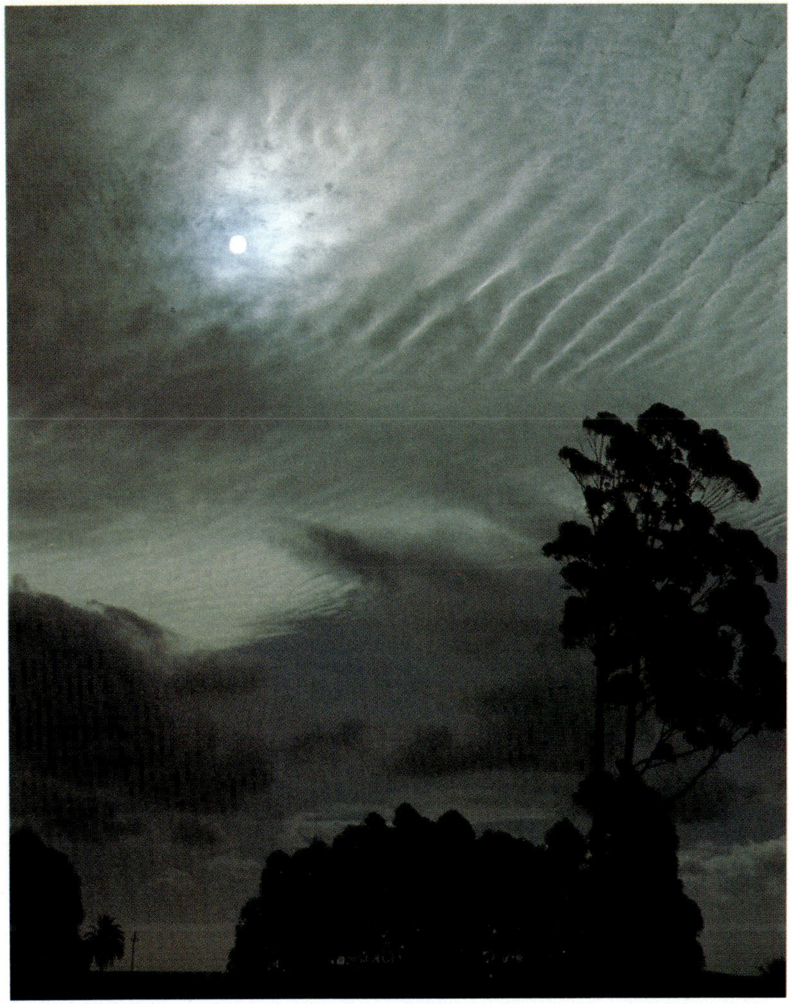

'The rain liquid it is which comes from the clouds . . .'

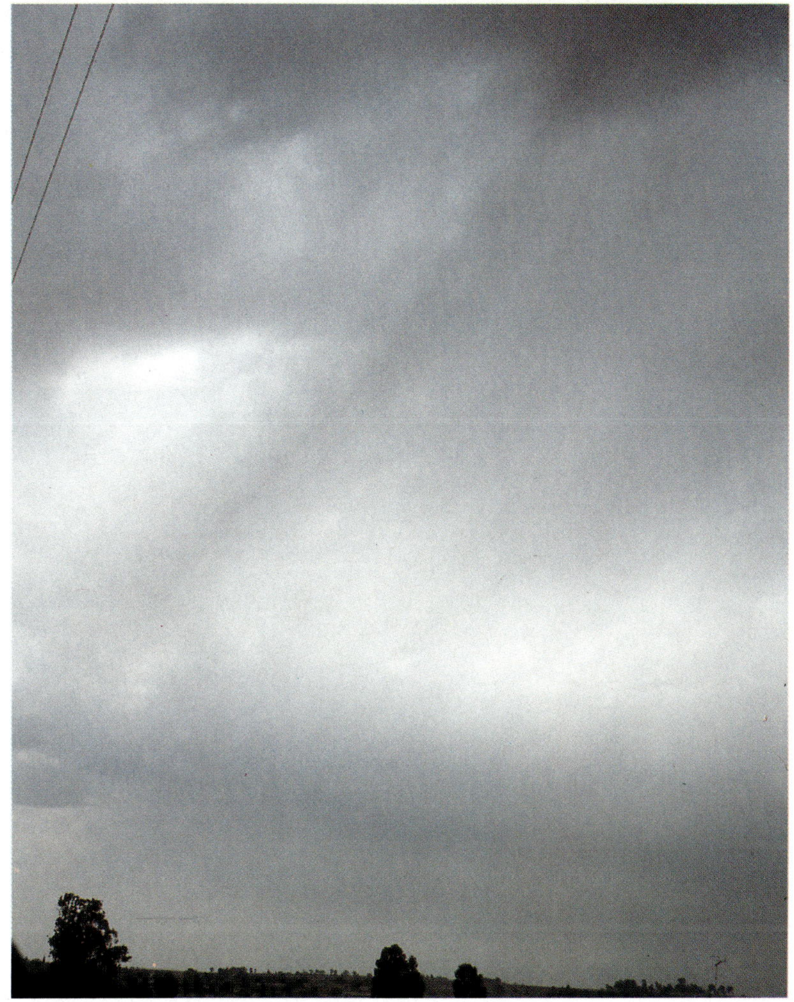

'It is that which people call the rain's leg'.

generally applied to persons involved in religious practices which produce states of trance with visions and out-of-body journeys of the spirit. Shamans profess to influence the good and evil spirits by their magic rituals.

Two other terms used in this book with which not all readers may be familiar are ales (plural alites) and therianthrope. Ales is a term adopted by rock-art researcher Harald Pager on the advice of Professor Sigurd Gangl of Graz to denote a painting of a winged or flying creature, or a human holding his arms in a wing-like posture. A therianthrope refers to a painting of a part-human, part-animal creature, usually with the head and feet of an animal and the body and hands of a human. The commonest examples have antelope heads and hooves.

Lastly, the use of a capital 'D' for 'District' denotes a magisterial district.

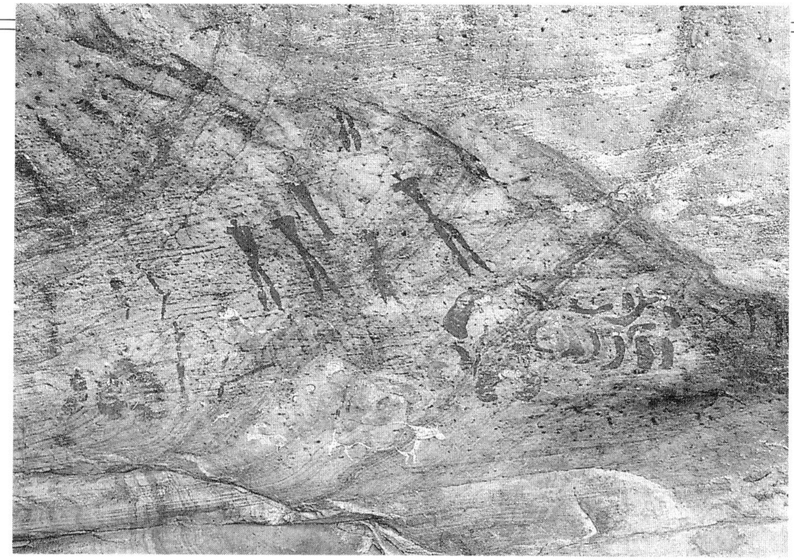

ACKNOWLEDGEMENTS

The material for this book has been collected over a period of some thirty years, much of it in the company of my wife Shirley and some of it with my children Diana, Lynne and John. Neil Lee has been my companion on most South African expeditions and we have been joined by our wives Shirley and Elaine on many occasions. On several trips our party has included, and been helped by, Robbie Steel and his wife Julia. The companionship and support of these, my family and close friends, is gratefully acknowledged. In Zimbabwe, too, I have always enjoyed the co-operation of members of the Zimbabwe Prehistory Society.

In other publications I have mentioned individual farmers who have been particularly helpful with directions, transport and hospitality. To all those and to countless others, together with officials of various parks boards, forestry stations, museums and libraries, I again record sincere appreciation and thanks for their assistance.

Special thanks are recorded to Fritz Pretorius of Slabberts, Orange Free State who, in recent years has made a farmhouse available for our use when doing fieldwork in the Caledon River catchment area. This work has been formalised under the auspices of the Institute for the Study of Man in Africa and has received generous financial assistance from the Anglo American and De Beers Chairman's Fund. Other members of ISMA, notably Sally Parker, have participated from time to time.

Opinions expressed in this book are my own unless specifically acknowledged.

Above: Personification of rain clouds? Clanwilliam.

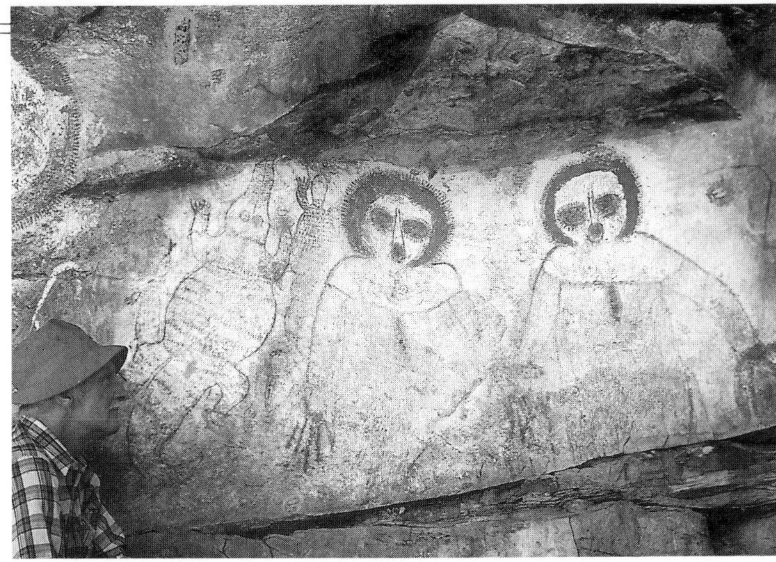

Figure 1

Rock Art, Man and Rain

For at least 30 000 years, and probably for much longer, people have been pecking, chipping and painting on the rocks of their environment; expressing their experiences, beliefs and even their aspirations. The resulting art, weathered and eroded by the passage of time, has lasted to the present day. Some of this art may have been casual and almost meaningless but a great deal of it is rich in meaning, imagery and symbolism. It was created by talented individuals who produced images which, in some cases, were illustrative and in others were probably considered to have the power to influence the behaviour of the natural forces over which man had no control.

Before the time when animals were domesticated and herded, before the days when crops were sown, cultivated and harvested, people lived by hunting and gathering. There was a broad division of labour: the men hunted, and the women gathered. Until recently, and even now in many parts of the world, men did not move without weapons. Women rarely moved without gathering bags. The men ensured the supply of meat for protein and skins for clothing. The women were responsible for the plant foods. The rain to bring on the growth of the vegetation was fundamental to both these activities and to the whole economic system. It is therefore no wonder that rainmaking, with all its associated magico-religious beliefs, was such a vital motivator for the art.

There is art on the rocks of every continent. It is to be found in the deep caves of Europe, among the canyons and mesas of America, on the steppes of Asia, and on the sandstone cliffs and krantzes of Australia and Africa. While much of this art is admired for its obvious beauty, a full appreciation requires some knowledge of the way of life of the artists and their contemporaries, and of the environment in which they painted or engraved. In Aus-

Above: The wandjina spirits of the Kimberleys, Western Australia.

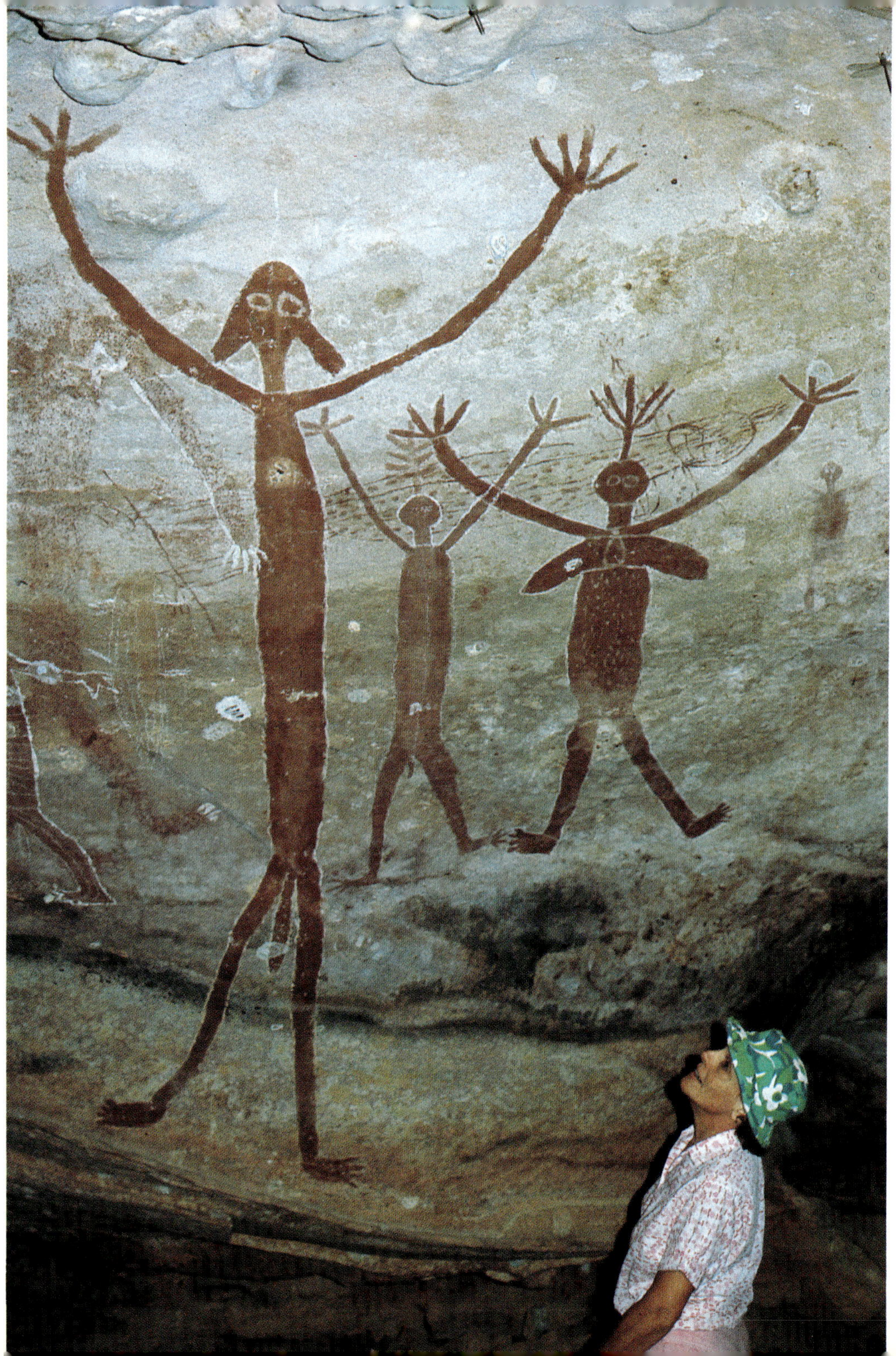

Figure 2: The quinkan spirits of Cape York with Beverley Trezise for scale. Her husband, Percy, put quinkan country on the map.

Figure 3: Turtle in X-ray. Oribi Rocks, Arnhemland.

Figure 4: The sacred spirals at New Grange, Ireland.

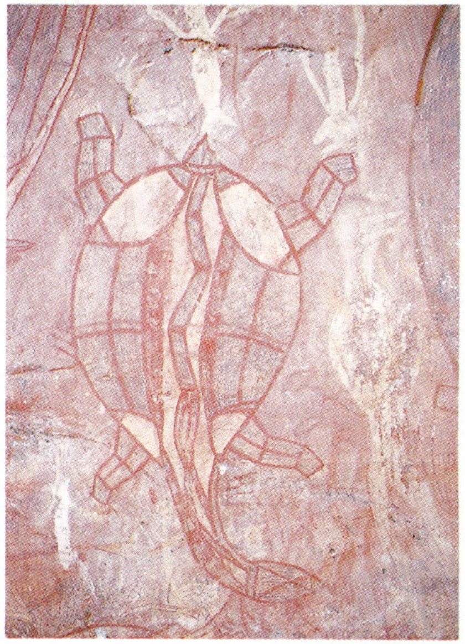

Figure 3

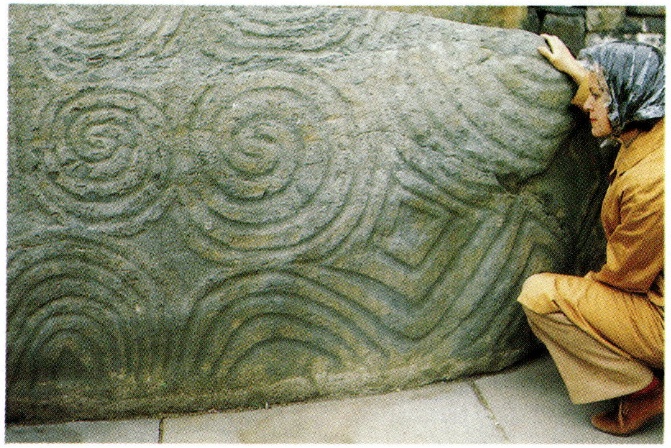

Figure 4

tralia, for instance, for thousands of years, communities of Aborigines were isolated from one another on the edges of a vast continent. Consequently, the engravings of sea creatures and animals on the sandstone ridges of the Ku-Rin-Gai Chase near Sydney are quite different from the engravings on the consolidated beach at Port Hedland in Western Australia. Although the Aborigines were all concerned with the spirits that influenced their lives, the quinkans, or spirits who emerged from between the rocks, of Cape York are not at all similar to the wandjinas, those spirits who arrived from the skies in the caves of the Kimberleys.

In spite of these differences in the art, however, the concept of a Dreamtime, when the land and its occupants were created and inextricably bound together in a sacred relationship, is fundamental to all aboriginal life. It is a concept that has a profound influence on relationships between the Aboriginal people and white Australians because of the clash between the sacred and the commercial, exemplified by the discovery of valuable minerals in what appears to be a barren, deserted piece of territory. This territory is nonetheless perceived by the Aborigines to be the home of an important Dreamtime character who was considered responsible for the creation of the minerals in the first place.

The Australians are particularly fortunate in the living heritage of indigenous peoples who can explain the meaning and significance of many of their paintings. The Institute of Aboriginal Studies in Canberra has amassed huge archives of stories, songs and films of dances and ceremonies, many of which relate directly to the paintings. Rainmaking and the journeys of the Rainbow Serpent feature prominently in these archives along with the Lightning Man and the Thunder Man. It is not surprising in a continent, where rain is likely to be distinguished by super-abundance or complete absence, that it provided an important theme for the art of both the painter and the storyteller.

A shift in focus from the hunter-gatherer way of life of the Australian Aborigines to the agrarian economy of, say California or Arizona in prehistoric times, nevertheless reveals the same concepts of rainmaking and of the rain snake but it also shows that there was greater awareness of the seasons and their relationship to the positions of easily observable heavenly bodies such as the sun, moon and prominent planets. The rock art sites are often situated on or alongside notable natural features such as standing stones or slits in rocks through which the positions of cast shadows can be marked by paintings to record the correct time for spring planting — a step forward from the magic of rainmaking to the science of weather forecasting.

In a comparatively wet country such as Norway or Sweden it is interesting to discover an art which shifted away from both rainmaking and the observation of the seasons to technology as represented by engravings of ships and two-wheeled carts. In the Valcamonica, of Northern Italy there was a further shift to land transportation and trade in the form of representations of metal artefacts and even a four-wheeled cart — as is logical in an area on the 'amber road' from the Baltic through the mountain barrier of the Alps to the civilised world of the Mediterranean.

It was probably from that world that the builders of megaliths — structures involving the use of huge stones — travelled by sea to the more northerly countries of Europe. These builders also produced 'art on the rocks' — but the rocks were their own building blocks — such as those at New Grange in Ireland where the themes are abstract. But the structures are also concerned with the seasonal positions of heavenly bodies and probably with the ceremonies enacted on important occasions such as solstices and equinoxes.

There were obviously rich rewards in power, prestige and presents waiting for the individual with the ability to influence the weather — to bring rain or to withhold it. At the same time there were people who were prepared to use their talents of observation and intelligence to further their ambitions by claiming the power to influence the weather and by devising procedures to produce rain. Such procedures were usually and wisely prolonged over an indefinite period which undoubtedly helped their efficacy! And reasons for the non-arrival of the rains were promptly given as a last resort if all else failed.

As herders and agriculturists developed more sophisticated techniques of agronomy and migrated to southern Africa, rainmaking became even more important. Without rain at the right time crops failed and hunger ensued. Many of the established rain doctors, medicine men of the rain or shamans among the hunter-gatherers continued their efforts for the benefit of the new economy. The rites that they had developed were practised in those places where they had always taken place — in the rock shelters or shallow caves where they lived or which had been set aside for this special purpose. It is on the surfaces of these rock shelters that many of the paintings tell the story of man's early efforts to do what he still aspires to do — to control the rain. What follows is an attempt to unravel that pictorial story and to pay tribute to others who have pursued a similar objective in one or other part of this far-flung and varied region.

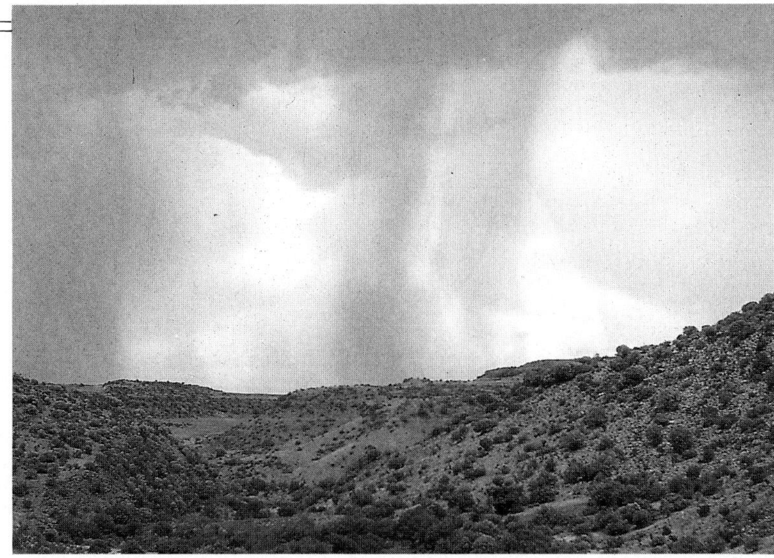

Figure 5

A Prototype Rain Animal: 'The Rain Has Legs'

A painting in the Bethlehem district provides an outstanding example of the Bushman concept of the rain animal (Figure 6). It was first published by Miss J R Harding in the *South African Archaeological Bulletin* in June 1950 (Harding, 1950) and her illustration was partially reproduced by Alex Willcox as one of a group designated 'rain animals' in *The Rock Art of South Africa* in 1963 (Willcox, 1963, 33). Neil Lee and I published a colour photograph in *Art on the Rocks of Southern Africa* in 1970 (Lee & Woodhouse, 1970) and I reproduced a black and white version in *The Bushman Art of Southern Africa* in 1979 (Woodhouse, 1979, 103). In 1978 I had drawn attention to the striking similarity between the falling rain in the painting and an actual shower of rain in the Orange Free State by publishing a photograph of the appropriate section of the painting alongside a photograph of a distant rainstorm (Woodhouse, 1978, 20) (Figure 5). So here is a rock painting with a history as well as a prehistory.

Harding submitted her copy to Professor van Riet Lowe for comment and he expressed the view that the painting was of hippopotami in pools of water with a stream running out of the right-hand pool. In the days before it became the practice to make a close comparison of rock art with relevant folklore and ethnography this was fair comment, but both the professor and Miss Harding were aware of some rather uncharacteristic features of the

Above: 'The rain has legs' said the Bushmen.

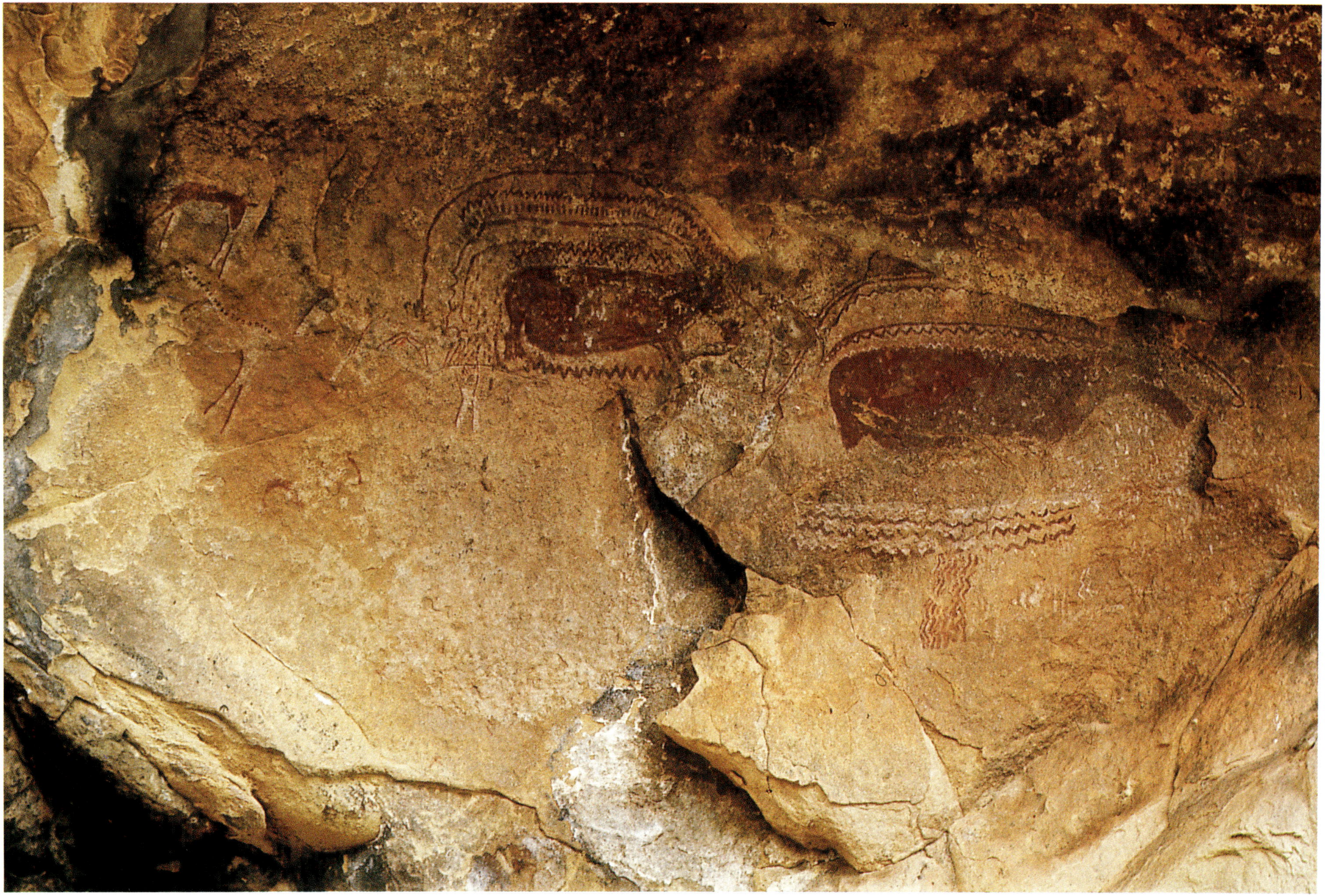

Figure 6: Prototype rain animals. The animals that lived in the rain clouds and sent the rain, according to the folklore of the Bushmen (San).

The zigzag pattern occurs with rain animals at a number of sites. There is a very faded white animal below this pattern. Clocolan.

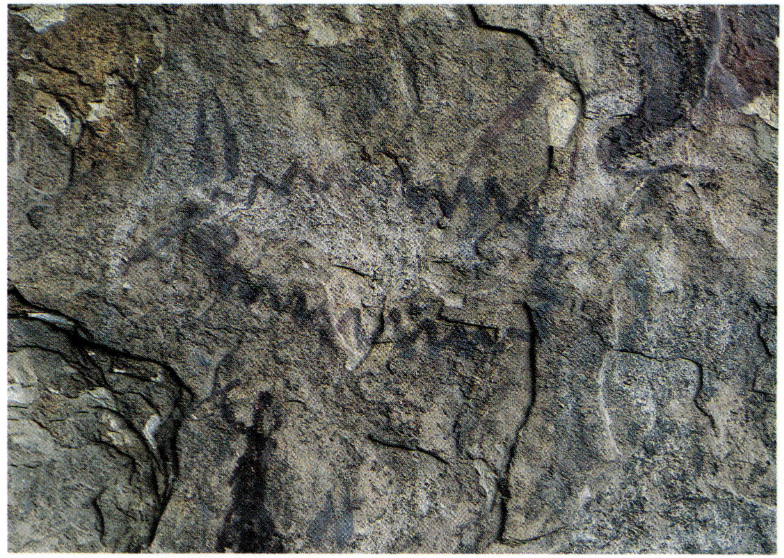

Figure 7

animals in the painting such as the narrow heads.

At a conference of the South African Association for the Advancement of Science in 1967, Neil Lee read a paper (Lee, 1968) in which he stressed the importance of reference to the relevant folklore in the study of rock art. Since that seminal paper we have learned, and are still learning, to look at the art through the eyes of the Bushmen. As a consequence we now have no difficulty in recognising that this painting illustrates the Bushman idea of a rain animal. It is conceptual rather than perceptual in essence but it certainly draws on well-known animals and rain showers as artists' models.

The similarity between many of the paintings of rain animals and hippopotami was commented on by Dr Sigrid Schmidt in 1979 (Schmidt, 1979) when she made the point that the rain bull often referred to in the Bushman folklore implied contact with cattle breeders; whereas the earlier paintings were probably created before such contact, when any thought of water would have included the hippopotamus which was abundant and widely distributed in rivers and vleis throughout the subcontinent.

Miss Harding wrote that the wavy lines in the painting reminded her of the chevron designs on the pottery of the BaSotho (who have lived in the vicinity for at least 200 years). Professor van Riet Lowe discounted this similarity by reminding her that 'the Cretans, Aztecs and others' also used the designs. 'They are universal symbols' he wrote. We shall, in fact, encounter them at widely distributed sites throughout southern Africa as we go in search of the rain and its creatures. The two views are complementary, not contradictory.

The rock shelter in which these paintings appear is small. Not more than five or six people can be under the overhang at the same time so it can be deduced that the paintings were not created to decorate a living place but to fulfil a special function connected with rain. There is, however, no record or tradition of rain-making ceremonies in the area and as yet no method of dating the paintings. Harding and Van Riet Lowe concurred that the animals to the left of the scene probably represented zebras but that they were rather stiff and might possibly have been painted by Bantu-speaking people (they used the term 'Bantu' which was still acceptable at that time). On the whole, however, they felt this was improbable and I would certainly agree. Stiffly painted animals do occur from time to time among the usually relaxed paintings of the Bushman and I take it that these 'zebras' are an integral part of the scene, for which the stiff posture was deliberately chosen. The lack of bold stripes (which are faithfully portrayed in paintings of zebra north of the Limpopo) raises some speculation as to whether these are not 'rain animals' of a different kind modelled on the

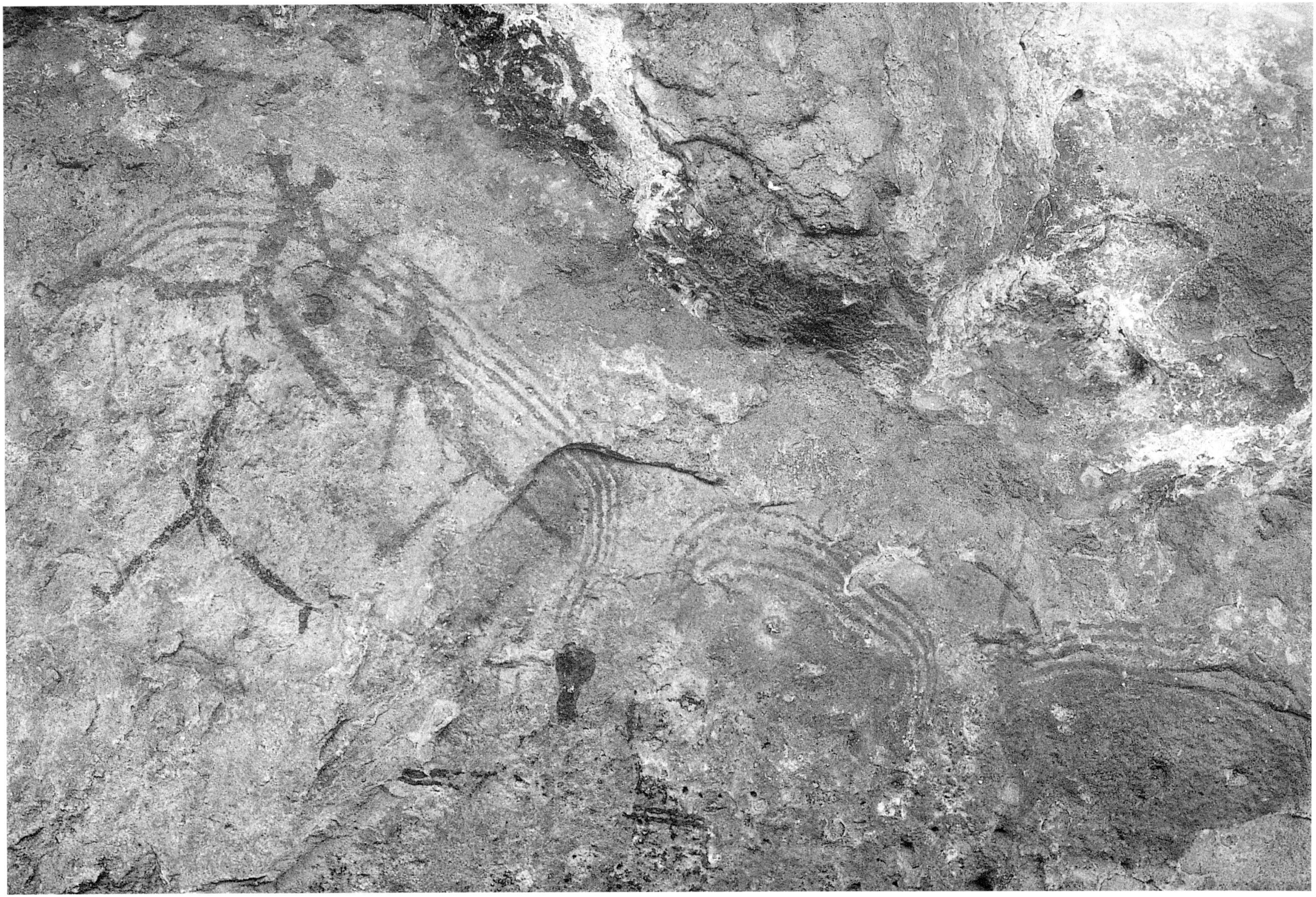

Figure 8: Rain animals in a similar style to the prototype from Bethlehem. The human figures were probably added subsequently. Fouriesburg.

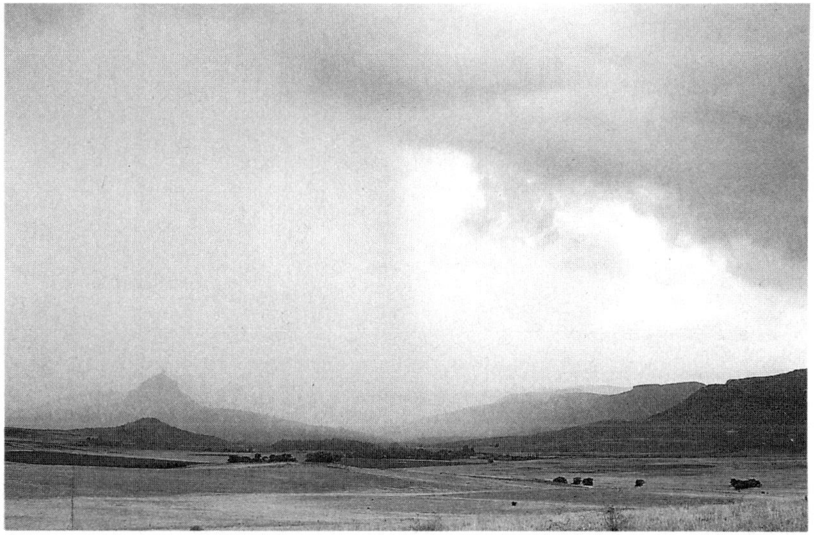

Figure 9: 'The rain has legs'. Fouriesburg, Orange Free State.

quagga with its lesser striping.

The accessibility of this scene indicates that it was created at a time of peace when the culture of the Later Stone Age was at its peak and when there had probably been some Early Iron Age arrivals who looked to the existing rain doctors for guidance on the local weather. Fortunately, the painting is not unique. There is a somewhat similar example a few kilometres away in the same area (Figure 8) and a variation which includes a faded white animal in the nearby Clocolan District (Figure 7).

There are many more interesting examples that lead us into some beautiful corners of southern Africa. They demonstrate the comparatively advanced nature of visual art in a culture that has so easily been dismissed as primitive — even barbaric — by early historians.

In support of the interpretation that in this painting there is a falling shower, is the Bushman observation that, 'The rain liquid it is which comes from the clouds. It is that which people call the rain's leg, when it does not rain everywhere.' (Bleek, 1933). As one drives across the Orange Free State on a summer's day it is highly likely that the legs of localised rain will be seen scattered round the horizon as a reminder of this observation and of the rain animals which the Bushmen imagined in the towering cumulonimbus clouds.

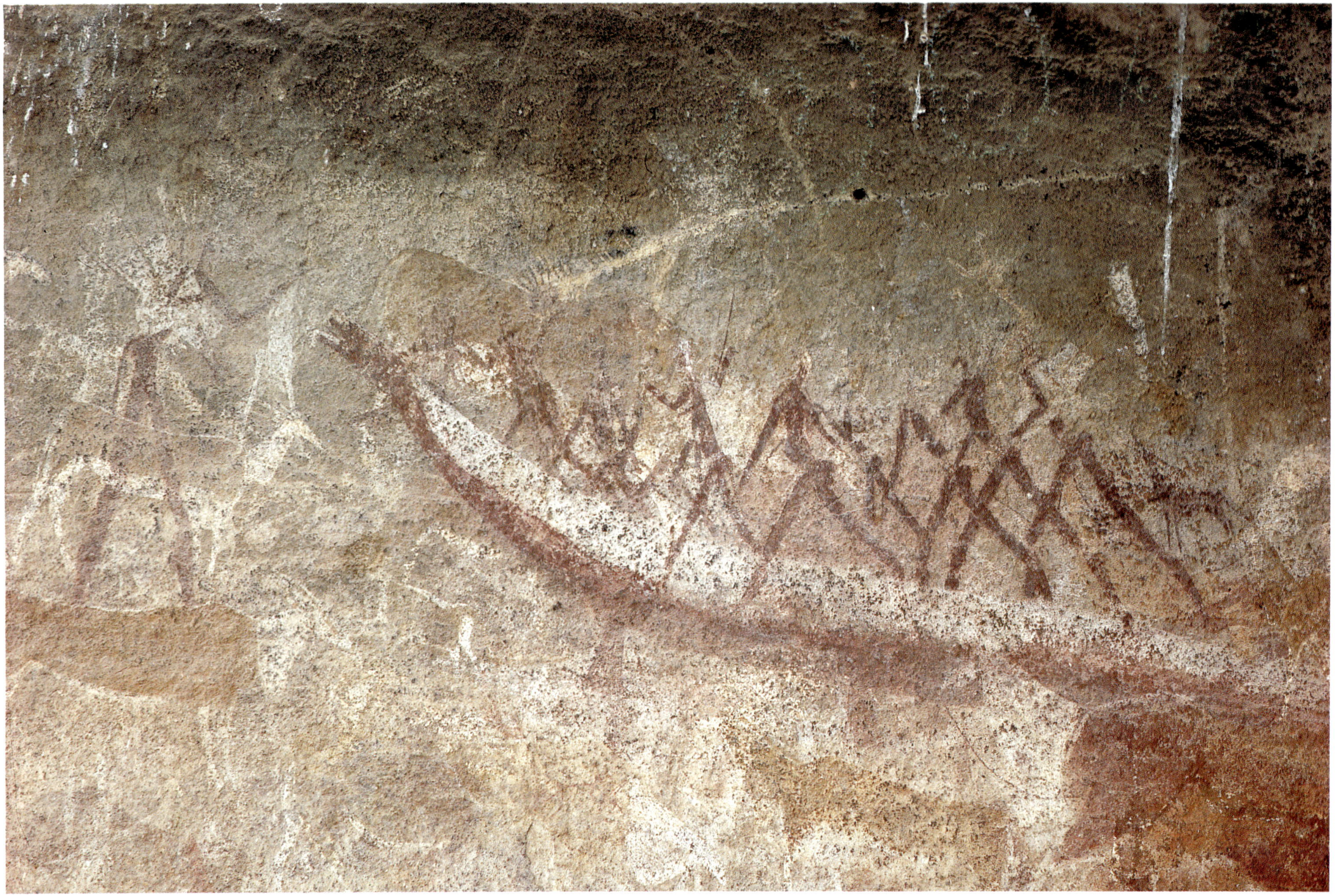

Figure 10: Offering of the corpse of a buck to a rain snake. A double-headed comet flashes above the snake and showers fall from it. Fouriesburg.

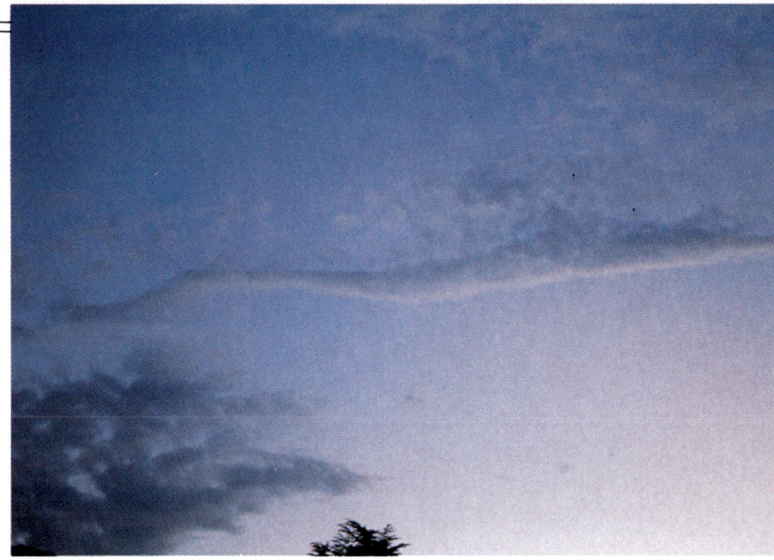

Figure 11

OFFERING TO A RAIN SNAKE

While the previously examined painting in the Bethlehem district is certainly sophisticated, it is a straightforward expression of the concepts of the rain animal and of the 'legs of rain'. In the adjacent district of Fouriesburg, however, there is a painting which raises a number of intriguing and challenging questions. Some answers will be attempted but the situation is open-ended because our knowledge of the non-material culture of the Later Stone Age and its overlap with the Early Iron Age is still incomplete. The discovery of other paintings may well augment an understanding in the future.

The main creature in this painting (Figure 10) is a snake raising its head and the front part of its body, which is not uncommon among the snakes of southern Africa. But this snake has two rain showers gushing from it in the form of closely drawn parallel lines. It is a rain snake. The Bushmen recognised a close association between some snakes and water, e.g. the puffadder whose green and yellow colours were compared with those of the rainbow. This, however, is a 'snake of the mind', possibly modelled on a python, and it is large in comparison to the human figures associated with it. There are at least eight who stride across it and another, larger than the rest, who faces it. From his outstretched and raised left hand hangs the limp carcase of a small buck. He is making an offering of this carcase to the rain snake.

Above the scene a double-headed comet (Woodhouse, 1986) flashes across from right to left. Such comets are known from historic records, e.g. Biela's Comet in 1846 and in 1852, but it is not suggested that the painting depicts that particular comet. In fact it seems extremely unlikely as it was probably painted much earlier. It does, however, indicate that two-headed comets were not unknown. Other features of the painting are a herd of eland, a smaller herd of mountain reedbuck and a man with arms raised,

Above: The origin of the rain snake?

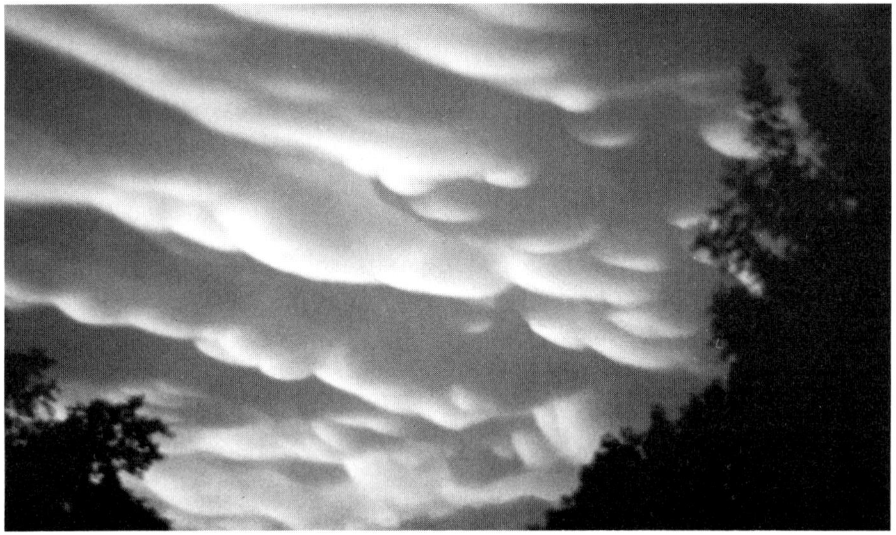

Figure 12

Mammalian cumulous cloud of the type that probably acted as a model for the prehistoric artists.

painted in pink. The man holding the carcase has a carrying bag containing bows and arrows, and several of the other figures are similarly equipped.

Frequent references by Bushman informants to snakes associated with water or rain (both expressed in the word *!khwa* (Lewis-Williams, 1977a)) and to rain showers falling from this snake — which are similar to the shower falling from the rain animal near Bethlehem only a day or two's walk away — confirm the essential concept of this painting but they do not throw any light on its 'sacrificial' aspect. In contrast, the practice of sacrifice is frequently recorded among the rain-making rituals of the Bantu-speaking peoples many of whom also associate snakes with rain (Schapera, 1971). The region near this painting was one in which there was contact between Bushmen and Bantu-speakers over a considerable period of time, so that the beliefs and customs of the latter were absorbed into local culture and are reflected in the paintings of the area.

The Bushmen were recognised by the Bantu-speakers as influential 'rain doctors' (Stanford, 1910) in the Drakensberg region and, in what is now Zimbabwe, several of their painted shelters became centres for rainmaking (Burkitt, 1928, 119). Consequently, there seems little doubt that the Fouriesburg painting was occasioned by, and may well illustrate, a rain-making ceremony. The addition of the herds of eland and mountain reedbuck are strongly indicative of the Bushman origin of the art, as are the overlapping legs of the striding figures which occur frequently in this region (Lee & Woodhouse, 1970, 58). The sacrificial element and the tallness of the figures may, however, indicate the influence of the Bantu-speaking clients. The addition of the two-headed comet portrays the sort of natural phenomenon that a quick-witted rain doctor would have turned to good account — either to explain the non-arrival of hoped-for rain or to strengthen and augment his own efforts to obtain it. The identity of the rain doctor in the painting is indefinite. He may be the man offering the dead buck or the pink figure with arms raised who could be regarded as orchestrating the proceedings. Or he may have chosen to remain anonymous.

The shelter itself is very secluded but it has a fine view of a breast- shaped hill which might be considered as highly appropriate to a place where the fertile blessing of rain was sought.

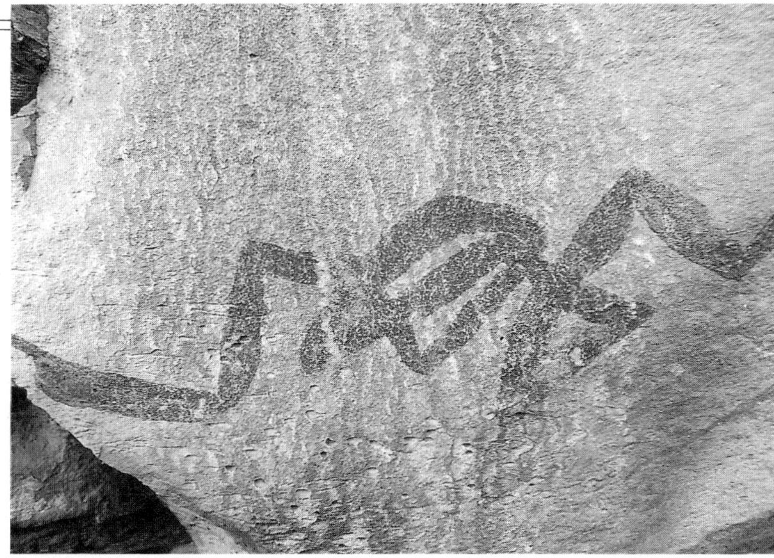

Figure 13

More Rain Snakes in the Eastern Orange Free State

If one knew where to look, when walking between the sandstone koppies and along the kranses of the scenic eastern Orange Free State, one would come to a painting of a rain snake, every day or two — or less. I write this with confidence because I have the photographic evidence, some of which is reproduced here. It would probably be the same story along the other side of the Lesotho border where my evidence is scantier but supportive.

Although each painting is different, most of them portray the same features of a person or people set against a background of short, vertical strokes, usually white, which are taken to depict rain. When a cold front moves across the Orange Free State it can easily be regarded as the arrival of the rain snake (Figure 11).

In the Rosendal district is the particularly clear, well-preserved example which was reproduced for the first time in my book *When Animals Were People* (Woodhouse, 1984, 90) (Figure 16). It is 1,3 metres long, which is longer than average for this part of southern Africa. The vertical white lines representing rain are painted above the snake but there are loops below

Above: Convoluted rain snake? Ladybrand.

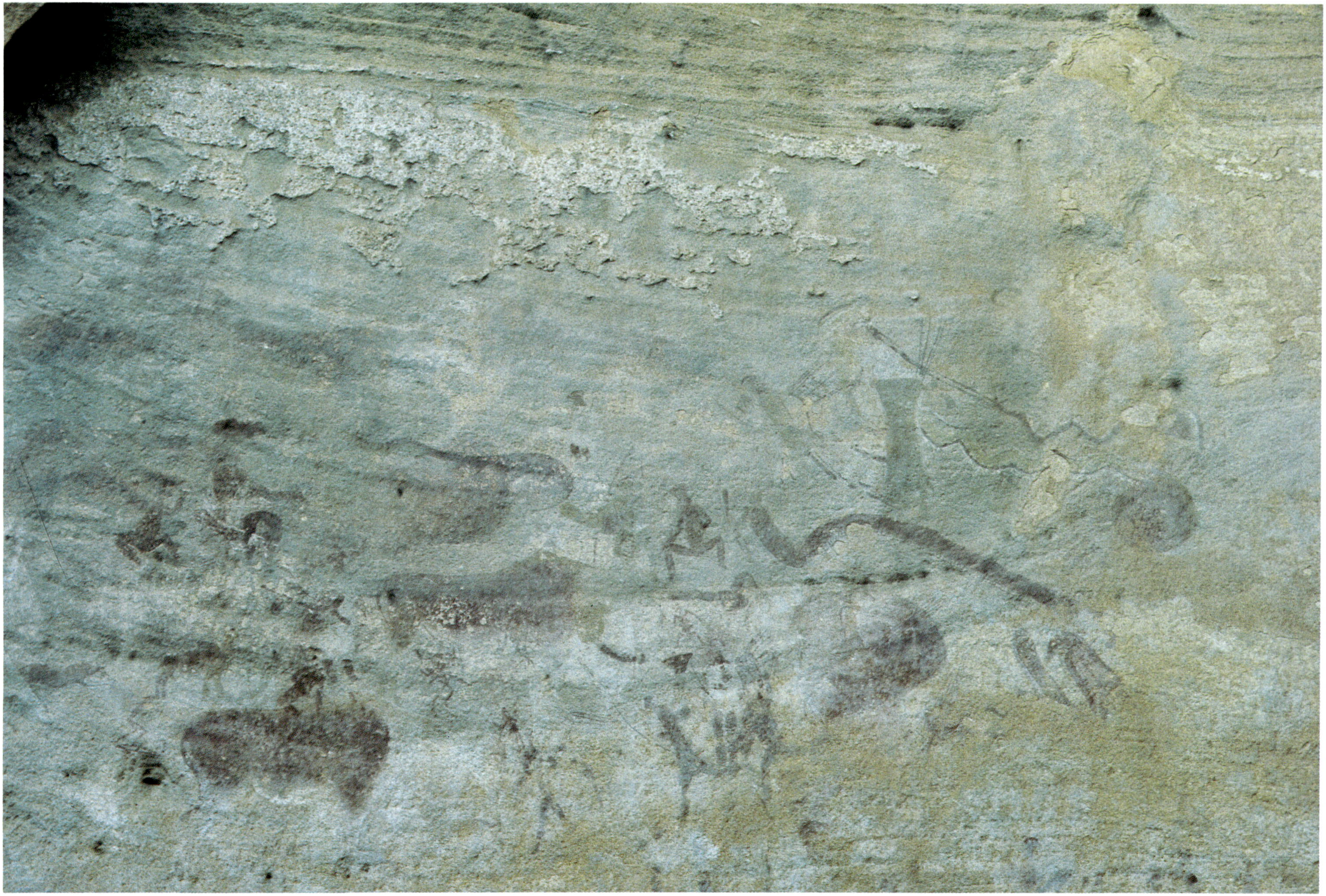

Figure 14: Three rain snakes on rock that has exfoliated, together with another comet. Ficksburg.

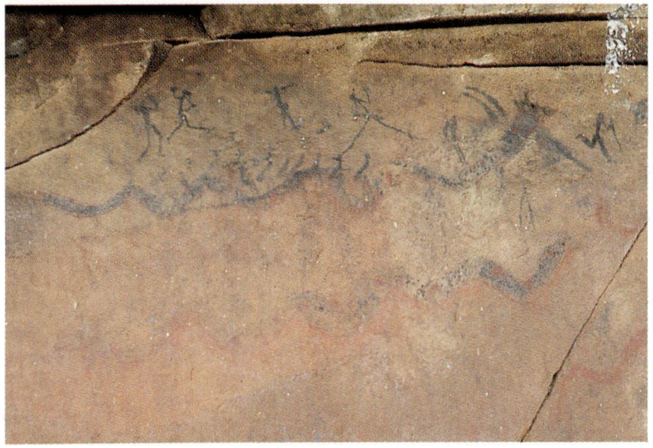

Figure 15: Two rain snakes originally copied and published by Walter Battiss.

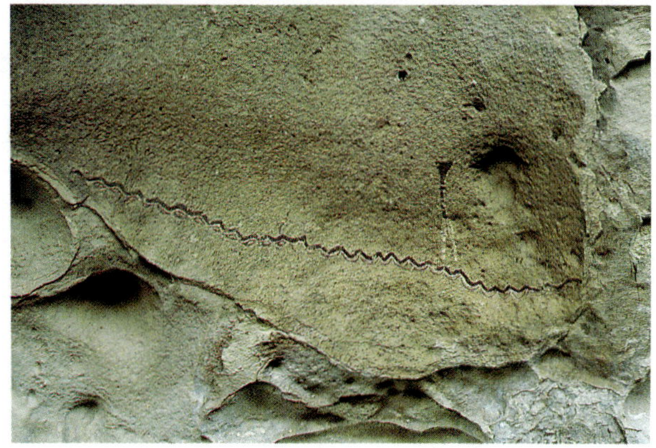

Figure 16: Rain snake with medicine-man of the rain? Ficksburg.

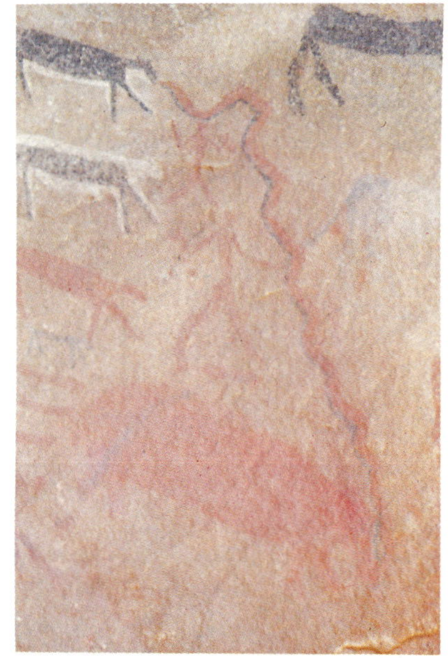

Figure 17: Rain snake and animals. Wepener.

which may be regarded as representing either clouds or rain caught by an updraft. The associated bichrome human figure has a wedge-shaped torso. This inactive but obviously significant character may be considered to represent the rain doctor.

From Rosendal to Ficksburg it is not unduly far to walk. In the Ficksburg district there is a small shelter which appears to contain three rain snakes and another comet (Figure 14). The paintings are extensively exfoliated — hence the reservations about identification. Yet this was apparently an important rain site where the three versions of the rain snake were painted separately over a period of time. The first snake is in the top right-hand corner of the frieze, painted in dark red outline with white rain streaks and yellow infill on the body. Immediately below it, is a large disc of dark red with thin lines radiating from it which I construe as the head of a comet whose tail stretches away to the left. This tail is white outlined in dark red, some of which has exfoliated. It thickens towards the end as the tails of some comets have been observed to do. To the left of the head of the comet is a man wearing a kaross and carrying a bow and arrows over his shoulder. He is one of a group scattered throughout the frieze. Below this figure is the second rain snake rearing towards a seated human. Both are painted in dark red monochrome with some white. The somewhat fragmentary remains of the third snake also rear towards a seated figure on the centre left of the frieze. Both the person and the snake are also painted in dark red and are spotted with white. A herd of eland and various human figures, together with the fragmentary remains of a pack of wild dog, complete the picture.

The snakes just described have much in common with one across the Caledon River in Lesotho (Figure 18). At that site is a snake with an elaborate head which was first illustrated by means of a drawing in *Art on the Rocks* (Lee & Woodhouse, 1970, 127). These two snakes are part of a frieze of a large herd of eland and there is the implication that copious rains will ensure plenty of grass, a bountiful increase in the eland herds and good hunting. One of the eland nearest the central snake is painted upside-down to signify its death from the efforts of the hunters — actual or hoped-for!

The most convoluted of the snakes of the eastern Orange Free State can be seen 'between bars' on the Ladybrand town lands where a small shelter has been fenced in as a precaution against vandals (Figure 13). The snake is again associated with eland including one painted upside-down. It was first reproduced by Helen Tongue (Tongue, 1909) who captioned it 'Snake or ornament'. Also in the Ladybrand district, close to the Caledon River are two snakes (Figure 15) which were reproduced by Walter Battiss in his classic work *The Artists of the Rocks* (Battiss, 1948, 209). Although not designated

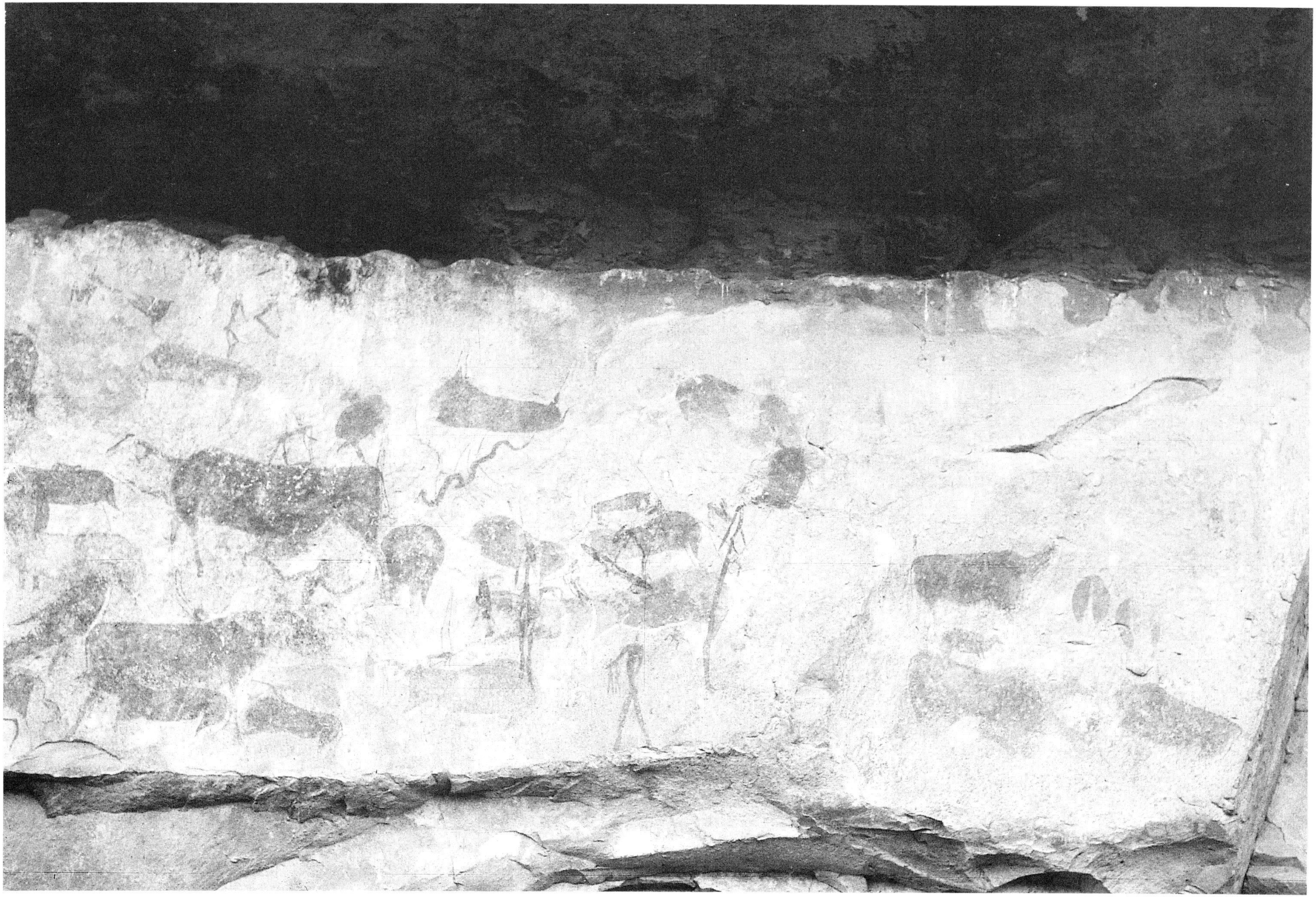

Figure 18: General view of the paintings in a rock shelter in Lesotho. The rain snake is left of centre. Above it is an upside-down eland.

'Rain animals gather'. Van Rhynsdorp.

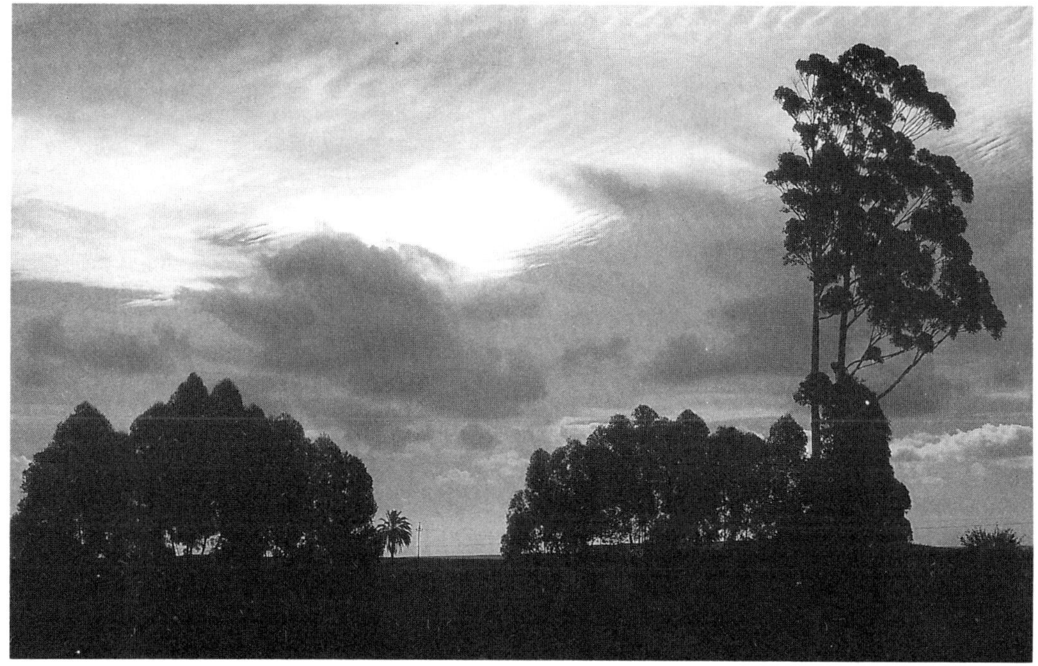

Figure 19

by Battiss as rain snakes (in fact his caption reads 'The horned snake is the famous mythical snake of the Basuto people and the second is a giant python'), there seems little doubt that both these snakes, together with a third painted below them, and the convoluted snake on the Ladybrand town lands, are part of the same story. An interesting feature of the Battiss snakes is the large red spots painted at more or less regular intervals along their bodies. They are associated with human figures painted in black and are superimposed on a large eland. Several of the human figures compose a dancing scene and two have their arms extended backwards in a significant position typical of the curing or trance dance of the Bushmen which is often depicted in their paintings, and which played an important part in Bushman culture.

A snake painted in the same style as the Ladybrand snakes, i.e. in red, edged with black, can be found at the national monument site near Wepener. Instead of being painted horizontally, however, it is almost vertical, (Figure 17) and is associated with domestic cattle.

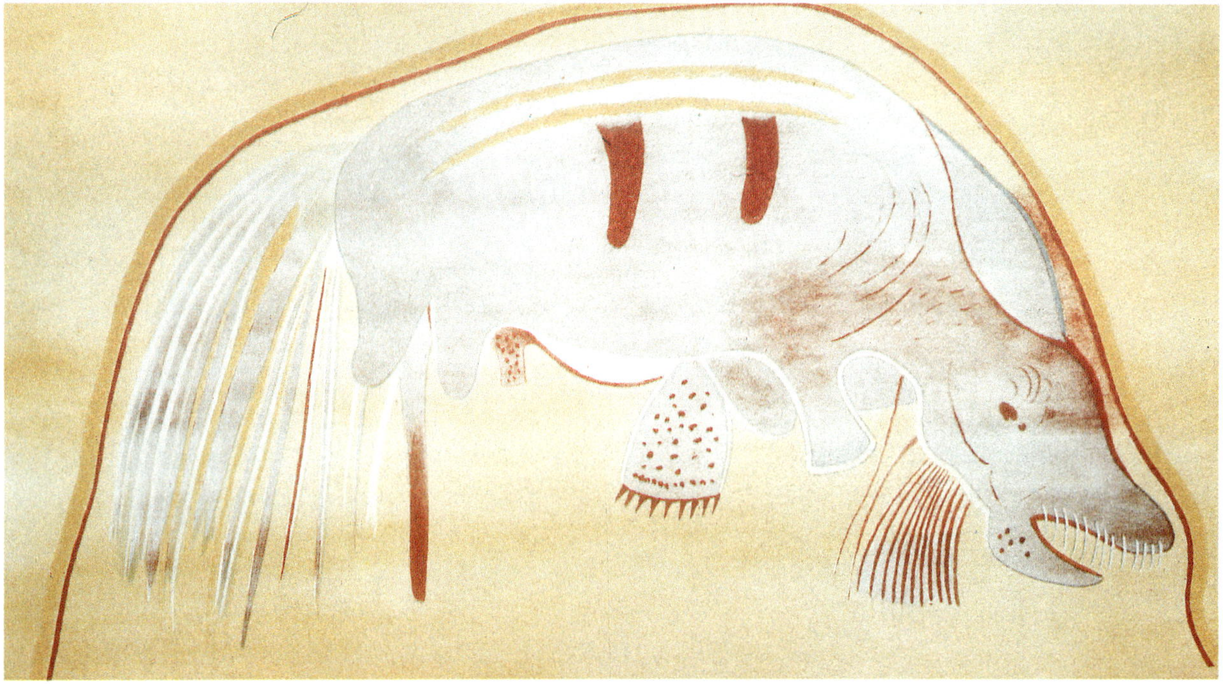

Figure 20:
Stow's copy of the 'she rain' made in the 1870s.

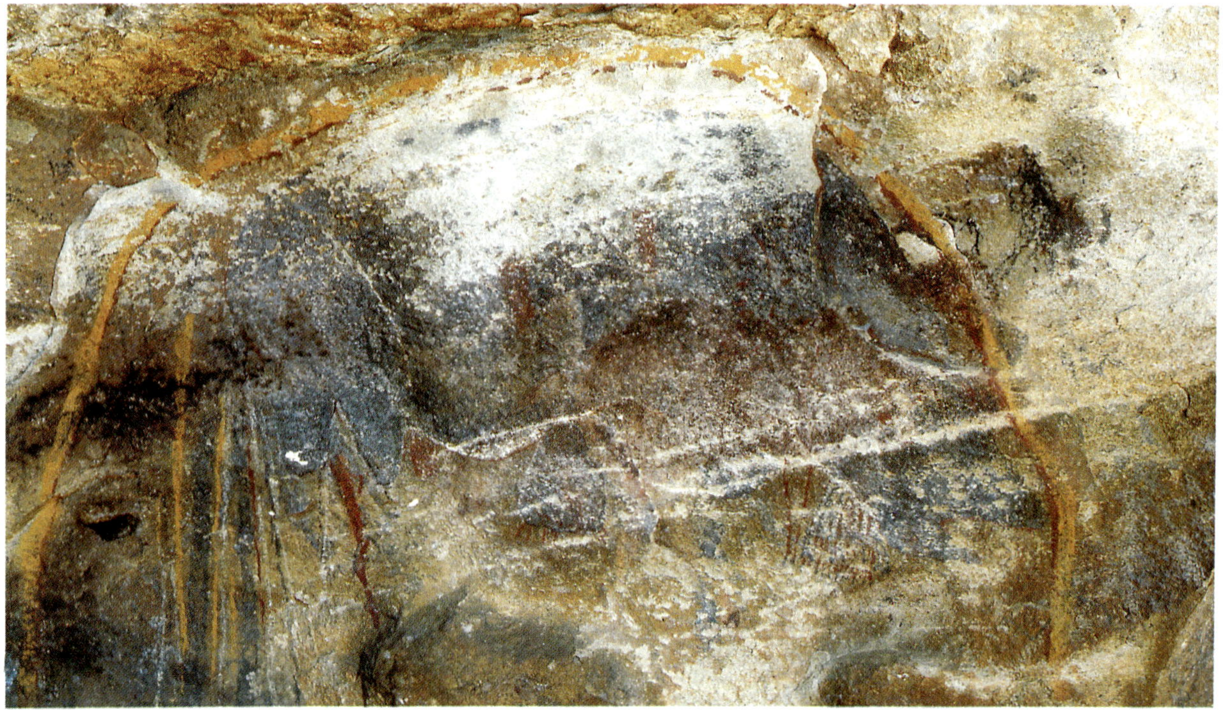

Figure 21:
Photograph of the 'she rain' taken in 1980. Rouxville.

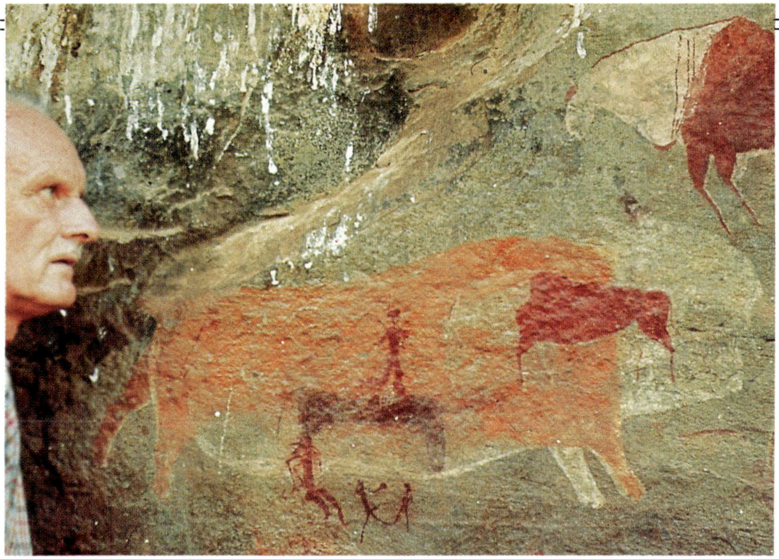

Figure 22

'HE-RAIN' AND 'SHE-RAIN'

When Dr W H I Bleek recorded the folklore and customs of the Bushmen in the 1870s one of his informants was *kabbo*, a Bushman from the Strontbergen in the Cape. Among many other things, he said that a grandson told his grandfather who was recognised as a rain doctor:

> You must not arouse a rain bull, but you must make a she-rain, which is not angry, which rains gently, because it is a slow shower. It is one that falls gently, softening the ground so that it may be wet inside the earth. For people are afraid of a he-rain, when they hear it come thundering, as it gets its legs.

When Lorna Marshall wrote of her research among the *!Kung* Bushmen in the Kalahari she wrote, 'After the earth *Gao!na* made the sky in a dome over the earth, and he made rain *!ga*. He made gentle female rain, *!gadi*, and violent male rain, *!ga!ga*.' (Marshall, 1962, 234). So the concept of male and female rain was widespread among the Bushmen. The main characteristic of the female rain is that it is 'gentle'. This characteristic is beautifully portrayed in the Clocolan District (Figure 23) where a gentle rain is squirted from the teats of a multi-breasted 'she-rain' animal. It has not previously been reported or reproduced, unlike a painting recorded by Stow at Klein Aasvogelkop in what is now an Orange Free State provincial nature reserve in the Rouxville district. He recorded it at more or less the same time as Dr Bleek was listening to *kabbo*.

Stow's copy was published in *Rock Paintings in South Africa* in 1930 (Stow & Bleek, 1930, 134) and it is kept in the South African Museum in

Above: A fierce 'he rain' animal with the author for scale. Photographed in 1971. Rouxville.

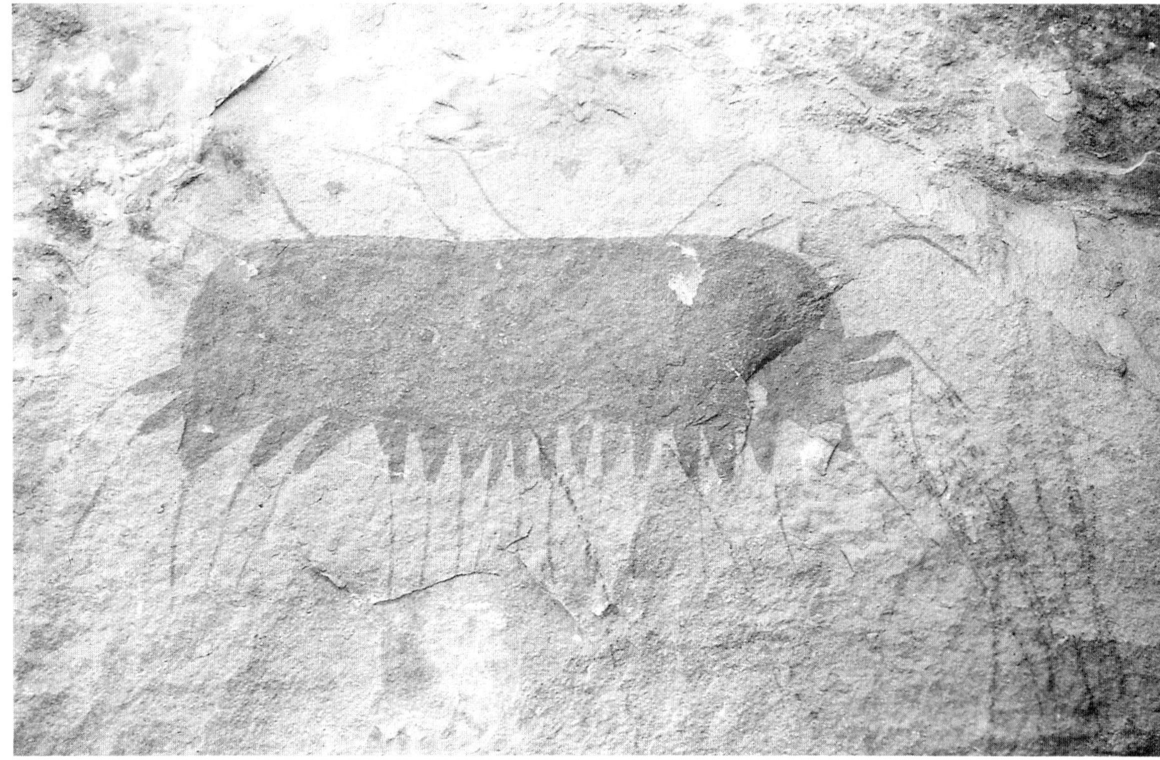

Figure 23:
The 'rain sow'. Clocolan.

Cape Town together with many others. It is reproduced here with a photograph of the actual painting as it looked in 1980 (Figures 20 and 21). The essential features are still discernible but exfoliation of the rock face has caused considerable deterioration and the body of the animal is black whereas Stow reproduced it as grey. It appears to have a rainbow arching over it — a phenomenon which will be dealt with in chapter 10 — which a Bushman informant told Dr Bleek was a distinguishing feature of a 'she-rain'.

Many examples outlined further on would seem to represent the sterner, more violent qualities of the 'he-rain'. A particularly fearsome specimen is reproduced here together with my face, as it was in 1976, to provide a scale (Figure 22). It typifies what George W Stow (1964, 112) describes as:

> . . . tremendous outbursts of elementary fury which sometimes visit the country, made their appearance, and their [the Bushmen's] superstition and dread were aroused, which among some of the tribes culminated in fits of impotent rage, as if the war of the elements excited their indignation against the mysterious power which they supposed was the cause of it.

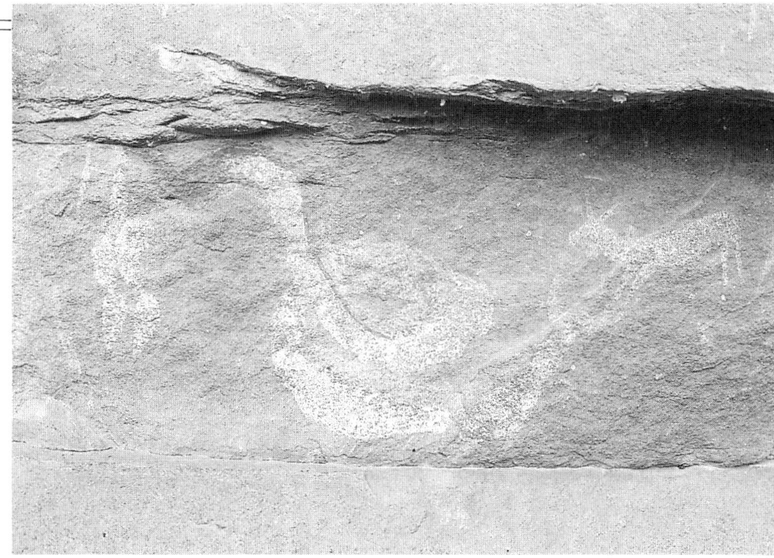

Figure 24

RAIN, FOOD CHAINS AND TRANCE

When exploring in search of Bushman paintings one has interesting experiences and meets charming people. One such occasion was in 1971 in the Rouxville district where, on our continuing 'walk-about' in the eastern Orange Free State, we found ourselves on the farm of the State President's brother. Neil Lee and I were received with warmth and hospitality — and were directed to some most interesting paintings. One of them is reproduced here (Figure 27).

The painting is a combination of rain snake and rain animal. Regrettably, the front half of the animal has been weathered away by water flowing down the rock face — it was obviously too efficacious for its own good!

Of particular interest is the unusual food chain illustrated in this painting. The usual food chain is rain — grass — antelope — man with bow and arrow — dead antelope. In this instance, the accompanying food chain is rain — edible plants — woman with digging stick — vegetables. There are also two birds which might have added eggs to the diet. Above, below and to the right of the woman and birds, are five dancing figures, two in black and three in red — the same colours as those used to paint the rain snake-animal. These dancers are certainly associated with the snake-animal, like the black dancing figures superimposed on the uppermost part of the Battiss snakes in the Ladybrand district. The red figure below has his arms thrust backwards in the significant position to which I first drew attention some years ago (Woodhouse, 1971), and which has become recognised as a symbol of the 'curing or trance dance' when the medicine men, a large proportion of the mature males, go into a trance and generate a release of tensions from among their small, closely-knit community (Marshall, 1969). Such

Above: Woman holding digging stick accompanied by rain snake. Symbolic of a food chain. Aliwal North.

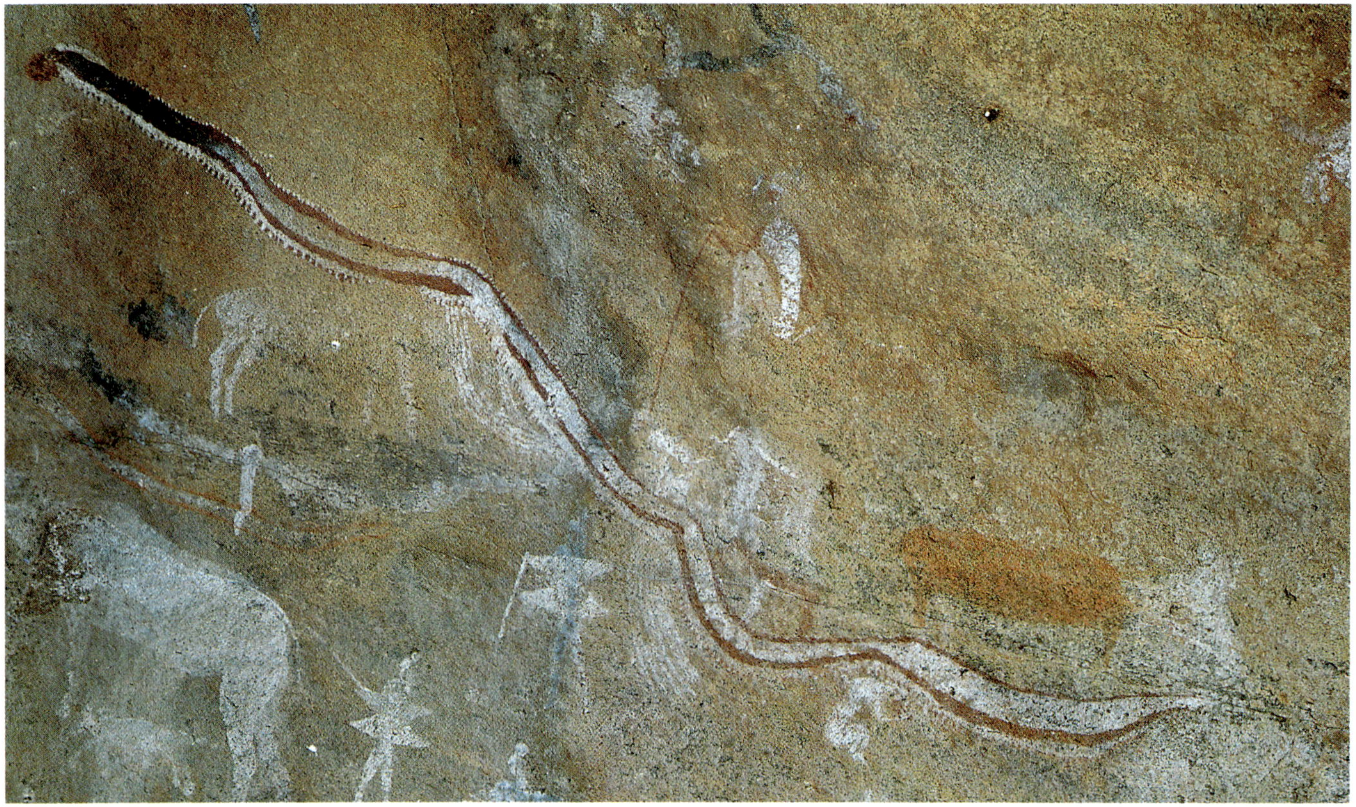

Figure 25: Rain snake with looping showers. Lady Grey.

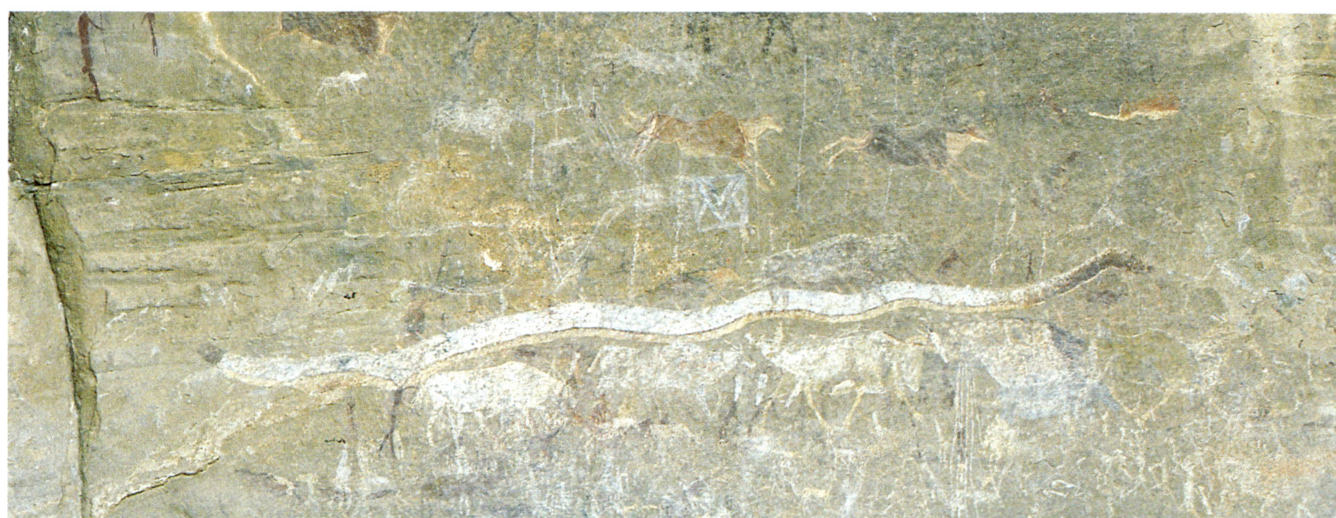

Figure 26: Rain snake confronted on right by man with assegai. A rock shelter in Aliwal North.

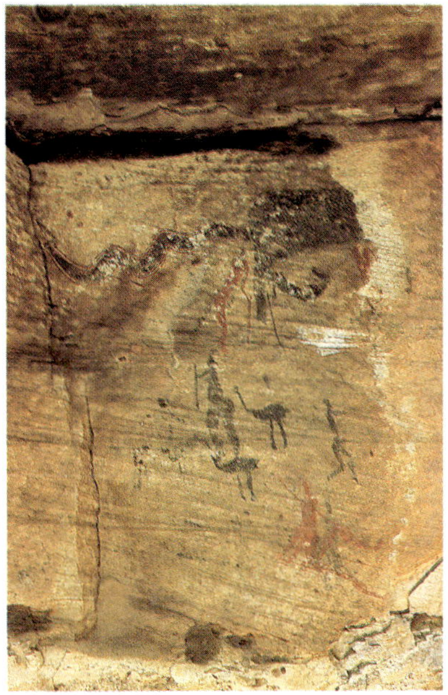

Figure 27: Rain animal/snake at top with possible food chain below. Rouxville.

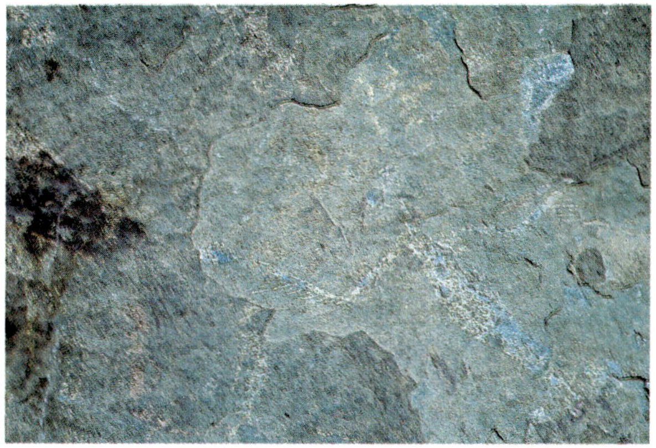

Figure 28: Detail of confrontation between rain snake and man with assegai. Aliwal North.

dances, or similar ones, would have taken place on the occasion of an important ceremony such as rainmaking, in which the actual painting of the rain animal and its accompanying dancing figures might have played a part.

The theme of snake and woman with digging stick — but with a fish and a buck instead of birds — is repeated in the Aliwal North district, but with the amply proportioned lady painted in white instead of black (Figure 24). In entering the Aliwal North district we have crossed the Orange River and entered the Cape but the boundary is obviously comparatively modern. The culture reflected in the paintings was the same on both sides of the Caledon and continues to be the same on both sides of the Orange. There is a considerable concentration of paintings in the Herschel district which is where the two provinces and Lesotho meet but a substantial proportion of this district is now Transkei.

At the same site as the woman with the snake is another, much longer, snake (Figure 22). It must be approached over rocks that are somewhat precariously piled on each other as the result of a recent rockfall. When making this sort of venture I cannot help wondering about statistics and pondering on whether I am destined to be around when a shelter collapses and whether, like Kaggen, the trickster hero of the Bushmen, I shall be able to summon my friends to dig me out. His friends, however, were the elephant, the rhinoceros and the giraffe (Woodhouse, 1984). Mine might not have the same muscular resources.

The head of the snake is threatened by a man with an assegai. There is a good tracing of this painting in *Images of Power* (Lewis-Williams & Dowson, 1989, 130). Red lines indicate that both snake and man are bleeding from the nose (Figure 28), and there are white streaks below the man's arms. His bleeding nose indicates that he is in trance, and the lines from under his arms probably indicate the falling of sweat — considered by the Bushmen to have great power. But they might also be rain. White lines which probably represent a heavy shower of rain fall from a rather lumpy 'animal' immediately below the snake, and there is an extensive panoply of hunters, dancers and lively eland both above and below the snake. The whole scene is patently a visual metaphor for rainmaking, but the threatening of the snake with an assegai is a feature of special interest as many other paintings in the vicinity depict the use of this weapon, rather than the more traditional bow and arrow — although the latter is not completely excluded. Art kept pace with technology.

In the nearby Rouxville District (Figure 89) there is a painting of a red snake with horns, above a lion-like animal with black horns, a seated man, a rather rotund animal and, to the left, a snake or slug-like creature together

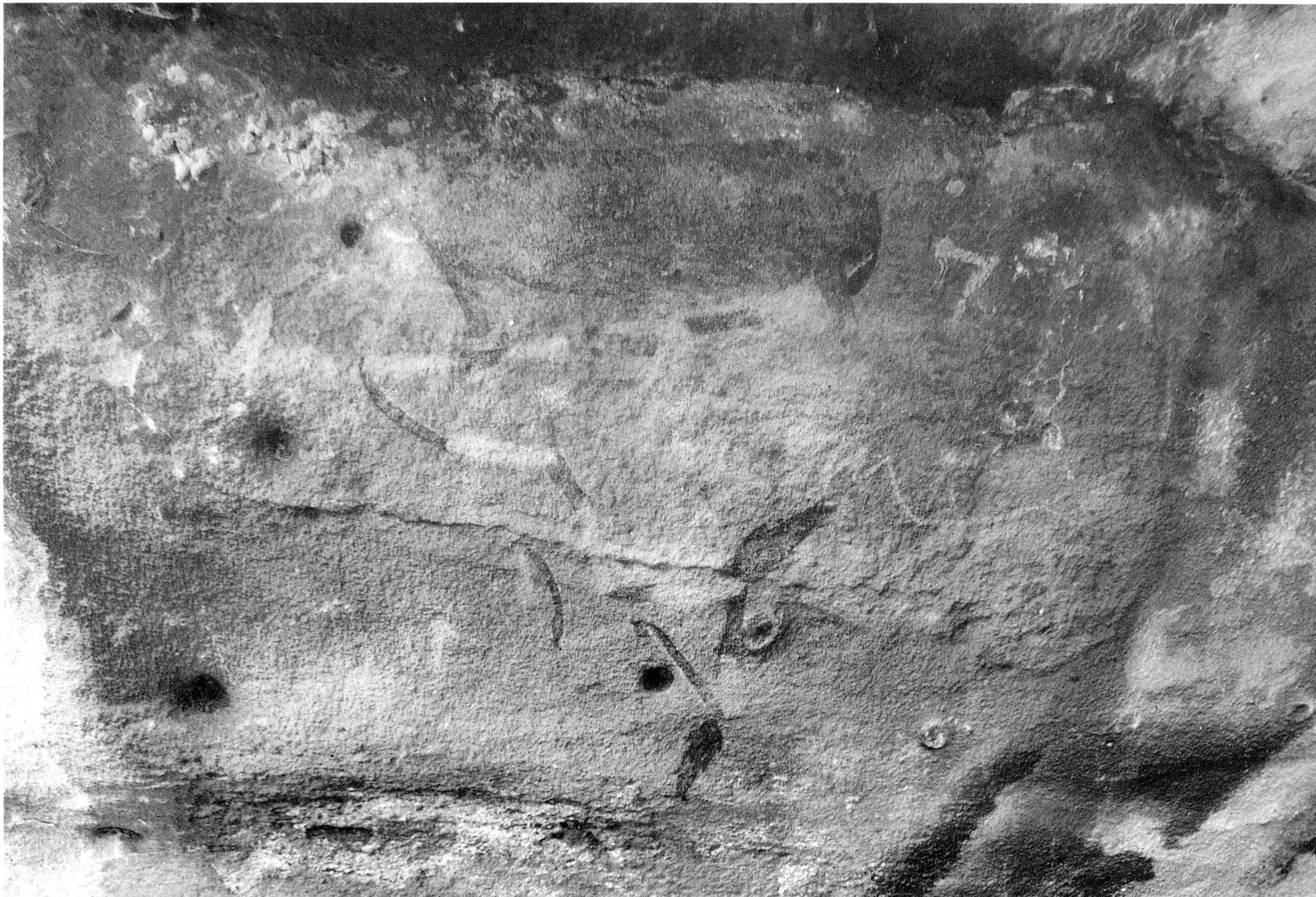

Figure 29: Trance experience and the rain animal. Lady Grey.

Figure 30:
Rain animal as river source? Lady Grey.

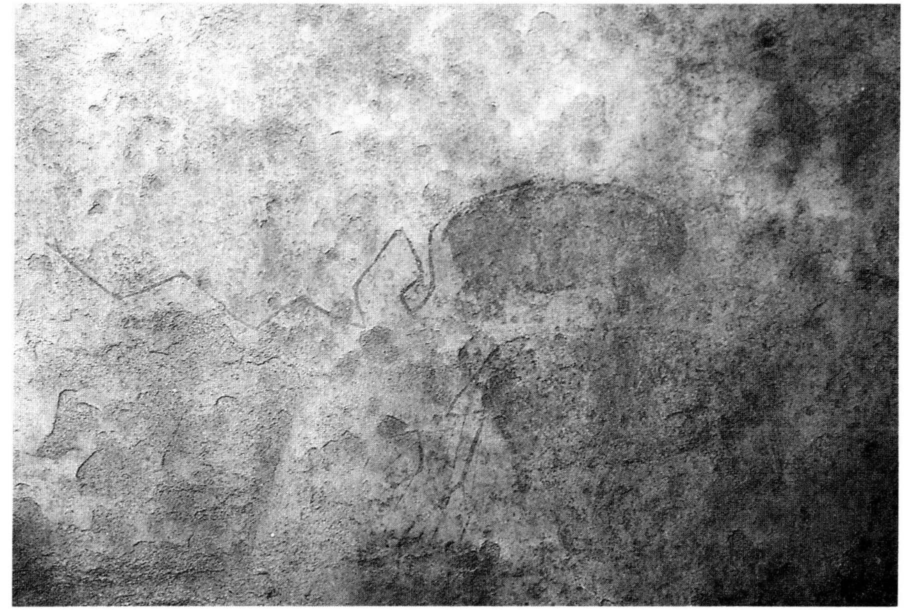

with an assortment of black dots, people and animals. The scene was copied by Stow (Stow & Bleek, 1930, plate 39) in the 1870s and revisited by Dorothea F Bleek in 1928, some fifty years later. I photographed it after another fifty years in 1977. Certain features are not as clear as in Stow's copy but, on the whole, the original seems to be surviving well in spite of some vandalism. There are a few minor colour errors in Stow's original. Dorothea Bleek commented that she considered the red animals to be mythical creatures and the men next to them to be medicine men. I would not disagree and would also consider it to be a rain-making scene. She compared the rather rotund animal with one at another site in the district (Figure 34), describing it as a 'spiked beast' — which it certainly is in Stow's copy. There are not many 'spikes' depicted in my photograph but white is a rather fugitive colour. She mentions that there is a copy of the second painting in the National Museum in Bloemfontein. This painting is of interest as it has suffered some damage and the copy might have been made before that occurred. It will be observed that, in spite of the damage, it can be assumed that a wavy line originates somewhere near the head of the animal and winds underneath it. There are several other examples of similar lines which will be dealt with further on.

Somewhat similar to the Rouxville painting is one in the Lady Grey district which was first reproduced by Walter Battiss and described as a 'decorated mythical snake' (Battiss, 1948, 208) (Figure 25). The 'decorations', white loops and curved lines below the body of the snake would appear to represent rain and there is a 'presiding rain doctor' above the snake and a shape rather similar to the slug-like creature below and to the left. An additional feature is the men with Sotho-type shields. There are many such figures at the Rouxville site — although they are not included in the rain-making scene. Shields of this type, together with the assegais, give some idea of dating — it is known that we are dealing with a time when Iron Age people had arrived in this part of southern Africa, and there was probably the sort of contact between tribes that was mentioned earlier. The white loops may be puzzling but, like more usual showers, they represent a meteorological phenomenon that may sometimes be observed when there is a strong wind below cloud level and the rain showers are deflected upward to evaporate before reaching the ground.

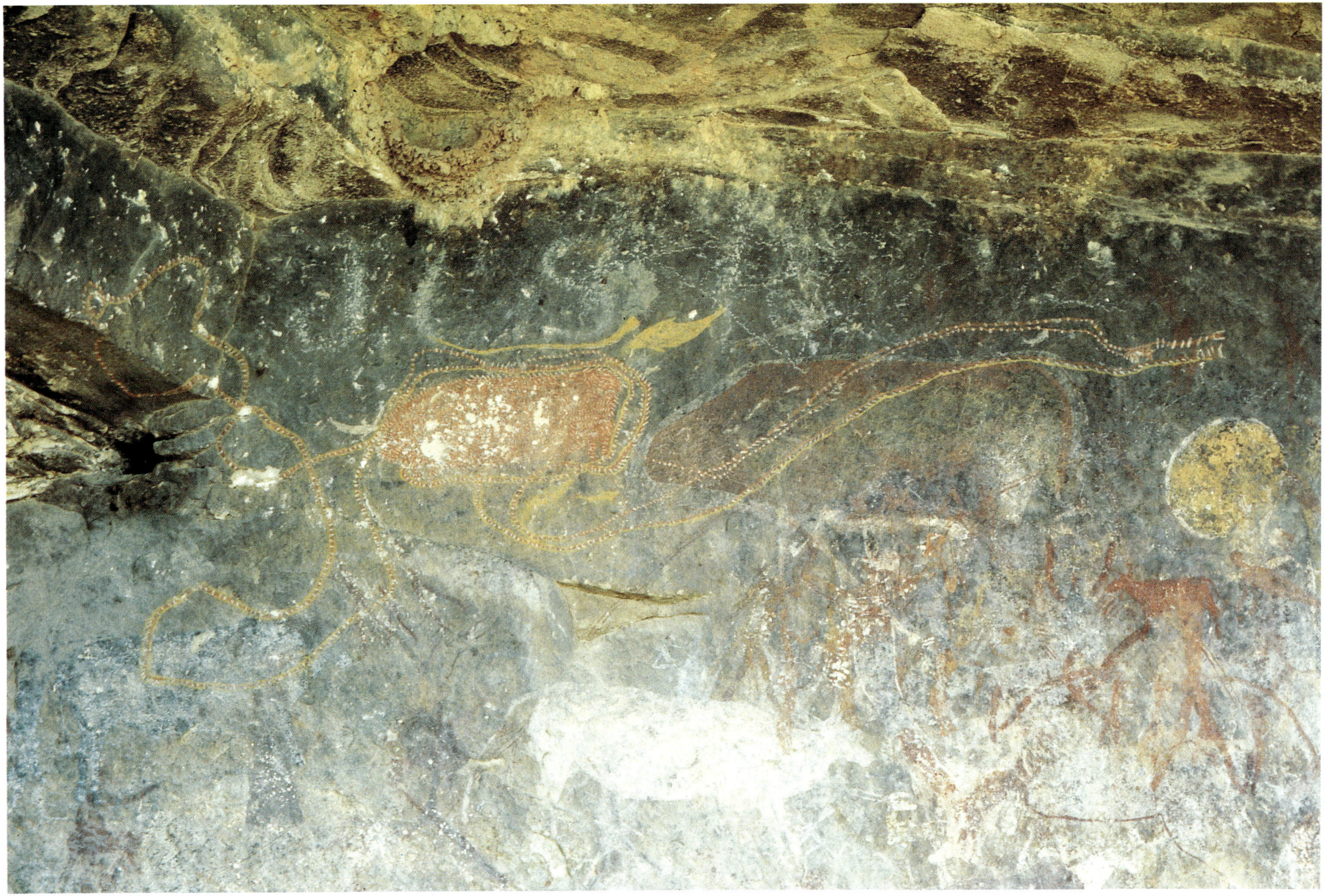

Figure 31: Sophisticated version of the rain animal/river source/fish concept. Herschel.

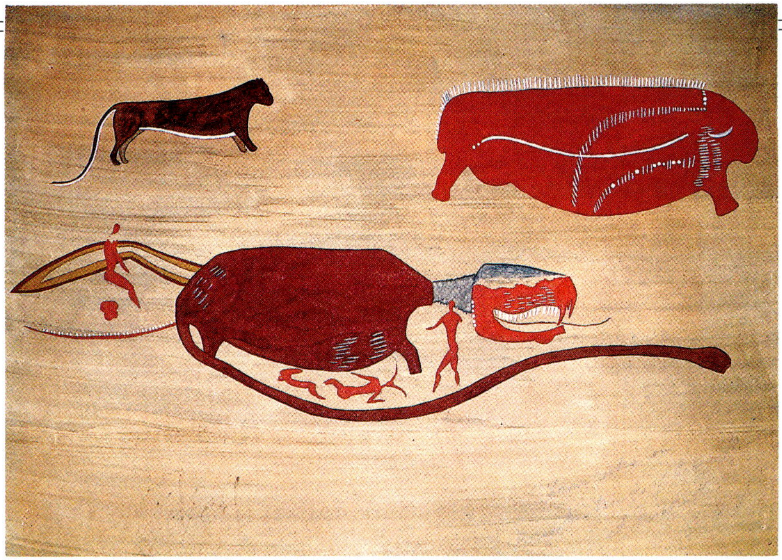

Figure 32

Rain Animal: River Source

Attention has been drawn to the wavy line round part of a rain animal depicted in a painting in the Rouxville District and to a fish in a rain snake scene near Aliwal North. These features are combined in an interesting painting from the Smithfield district which Neil Lee and I first published some years ago (Lee & Woodhouse, 1970, Figure 192) in which there is an animal almost, but not quite, resembling a hippopotamus from the head of which a line goes round the creature until it ends in front with a short right-angle turn (Figure 33). There are four red fish near the front of the animal and the remnants of others at the back which are painted in black and which were possibly added at another time. It has been suggested in the past that the right-angle at the end of the line may represent a gorge, used in fishing before the invention of hooks, for which there is supportive archaeological evidence. Below this creature are people, seated and standing, who are holding long, thin objects which appear to have heads and are likely to be eels. There are eels clearly depicted in other paintings, e.g. one from the farm Linton, Maclear, now kept in the South African Museum, Cape Town.

Two paintings similar to that near Smithfield were published at the same time. They are found close together in a shelter in the Herschel district (One is figure 35). In one painting the line emerging from the animal's head surrounds the animal; in the other it forms a loop below the head before going away from the animal. There are fish associated with the second painting but not the first.

At the same site as the handsome snake near Lady Grey are two more similar rain animals, but in the case of these two the line passes over their

Above: Stow's copy of a rain animal at Miaputti which has almost completely disappeared. Copy by permission of the South African Museum, Cape Town.

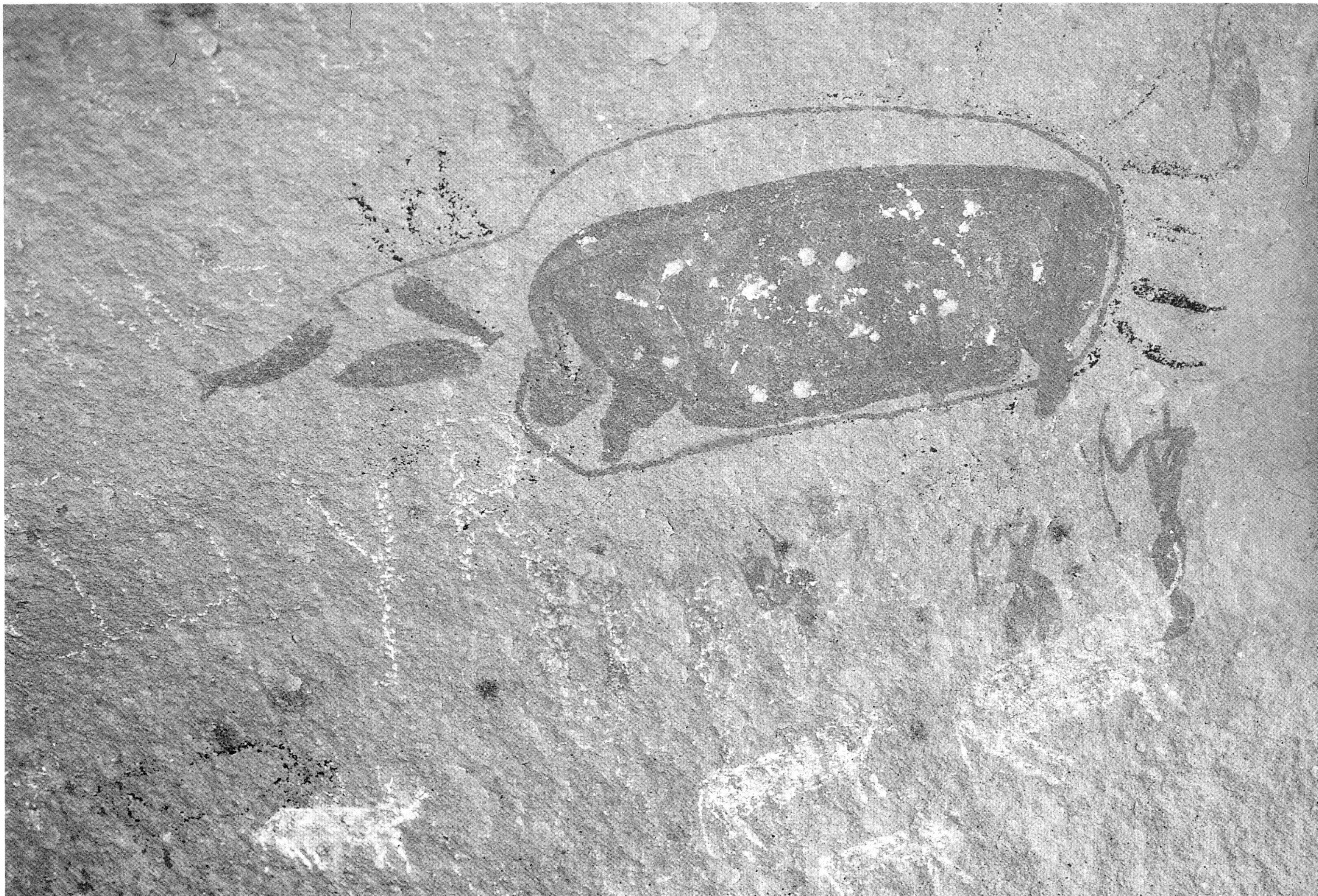

Figure 33: Rain animal with fish and eels? Smithfield.

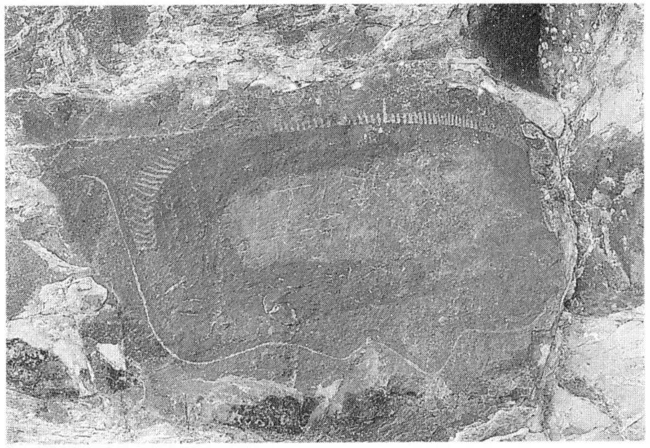

Figure 34: The rain animal as a river source? Rouxville.

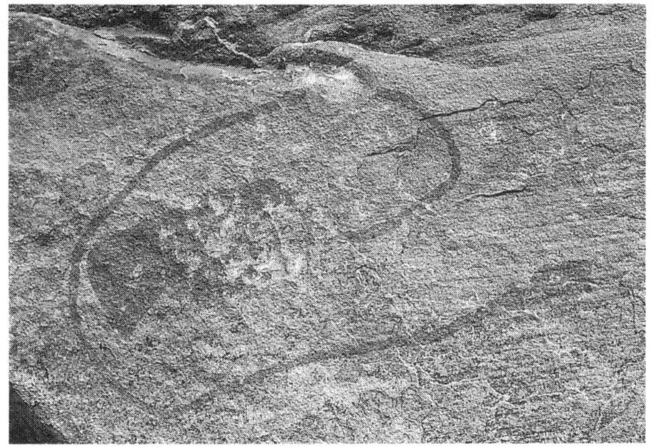

Figure 35: Rain animal as river source? Herschel.

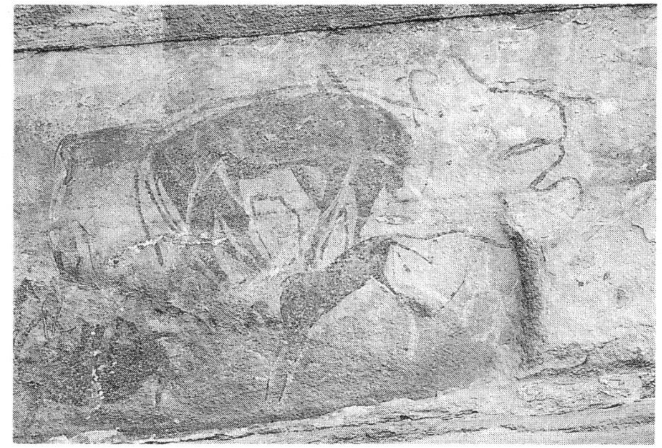

Figure 36: A 'monster of the water'. Wodehouse.

backs before assuming somewhat contorted shapes behind them (Figure 30). Not so far away from these rock paintings, near Sterkspruit is a comparatively well-preserved and complicated example (Figure 31) in which several lines, dotted with white, radiate from the rain animal, surround it and head off towards left and right directions. There are also yellow fish, a white fish, a long eel and another, rather lumpy-looking rain animal associated with the scene.

The cumulative evidence of these rain animals and connected lines — from Smithfield to Sterkspruit — seems to indicate the concept of 'the rain animal [being] the source of rivers — or streams of water — with consequent fish' — a logical concept in the circumstances. Another explanation has, however, been suggested by Lewis-Williams (1983, 50) based on an example in the Lady Grey district (Figure 29). He draws attention to the fact that certain of the human figures in the scene are bleeding from the nose while in trance and attempting to control the rain animal which is suggested as being near to death — denoted by the erect white hairs on its back. I have no difficulty in agreeing with this interpretation. He goes on to suggest that the fish underscore the similarity between trance experience and being under water. While I agree that this similarity may well be reflected in certain of the paintings, I take the view that in this case the fish are intended, as their positions imply, to be associated with the segmented line rather than with the man. None of these explanations takes special cognisance of the black and yellow segments, the significance of which awaits the illumination of some future discovery among similar paintings or the relevant ethnographic evidence.

A further example of the rain animal-river concept has been recorded by Harald Pager from a shelter near the Ndofela road in the Herschel district (Pager, 1975, 48), in close proximity to the same district where Stow recorded a fearsome example on the farm Miaputti. Regrettably, that monster is now so faded as to be barely recognisable (Rosenthal, 1953, plate 9) (Figure 32). There is, however, a somewhat similar creature, well-preserved, in the Wodehouse District (Figure 36) in a shelter which contains other rain-making scenes (Figures 42 and 43) which will be dealt with in subsequent chapters.

A scene containing fish or dolphins close to Caledon's Poort has been written about frequently (e.g. Woodhouse, 1984, 41-3) mainly with a view to identifying the fish dolphins. What has never previously been noted is the rain animal lurking in a recess on the left-hand side. It is undoubtedly part of the same composition and it reinforces the relationship of the whole scene to water (Figure 37) and therefore to rain.

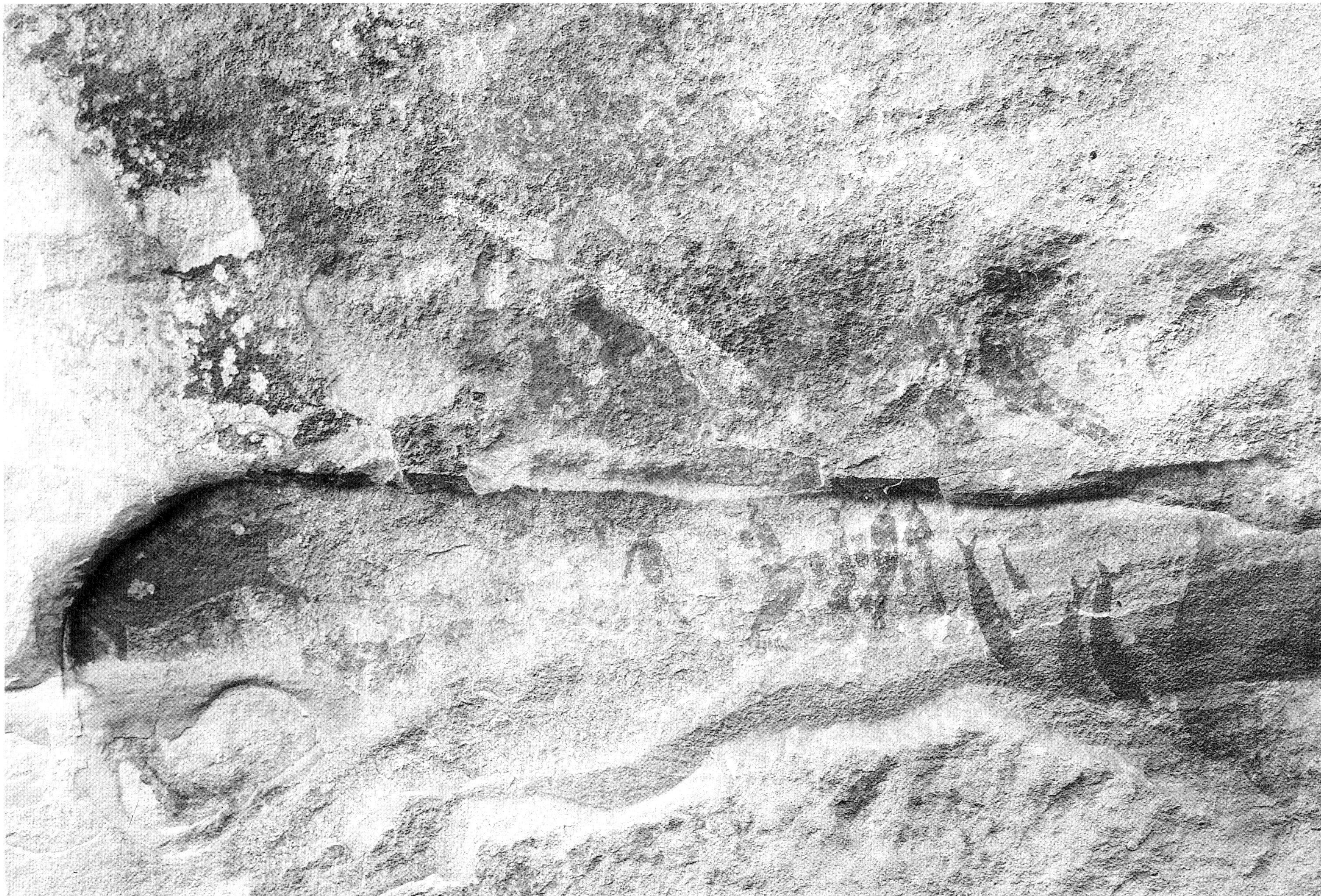

Figure 37: Fish or dolphins? With a rain animal under the recess on the left. Fouriesburg.

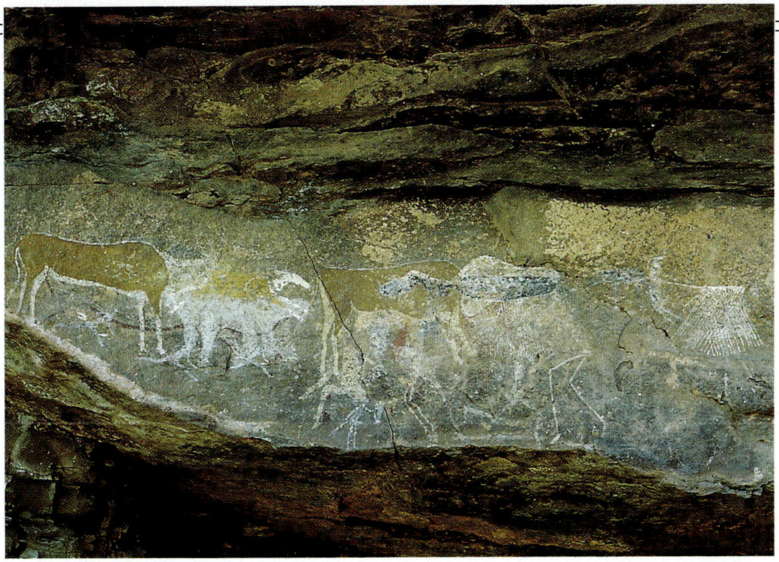

Figure 38

FALLING RAIN

Attention has been drawn to the rain shower falling from the cloud below the prototype rain animal in the third chapter. There are other examples. One, copied by Jalmar Rudner, in the Calvinia district appears to depict clouds with showers falling from them — the rain's legs — accompanied by two people, the taller of whom may be a rain doctor or shaman (Figure 39).

The painting was reproduced on the cover of issue number 54 of the *South African Archaeological Bulletin* in June 1959. There is a somewhat similar painting in the Wodehouse District with more scattered clouds, and another including possible flashes of lightning was copied in the Clanwilliam district by H Rabinowitz, Townley Johnson and P Sieff, and reproduced on the cover of issue number 59 of the *South African Archaeological Bulletin.*

Rain would appear to be falling heavily from a rain animal depicted in a painting in the Aliwal North district if it can be assumed that the artist had conventionalised his rain by the use of thin white parallel streaks (Figure 38). The same technique has been used at a remote site in the Indwe District (Figure 134) where a herd of eland is fleeing through an apparent rainstorm, driven by a creature with a human body but the head of a warthog (Figure 134 and dustcover). The brilliantly painted leading animals appear to be falling over a cliff — which was a customary method of hunting where the terrain was suitable.

The partly human, partly animal figures which appear frequently in the paintings have presented a challenge to interpretation since they were first remarked on in print by George William Stow (Stow, 1905, 98) who expressed the opinion that many represented *'Nadro,* a game of dressing up as an animal in which the chief performer was one of the older women. He did,

Above: The fine, white lines on right represent falling rain.

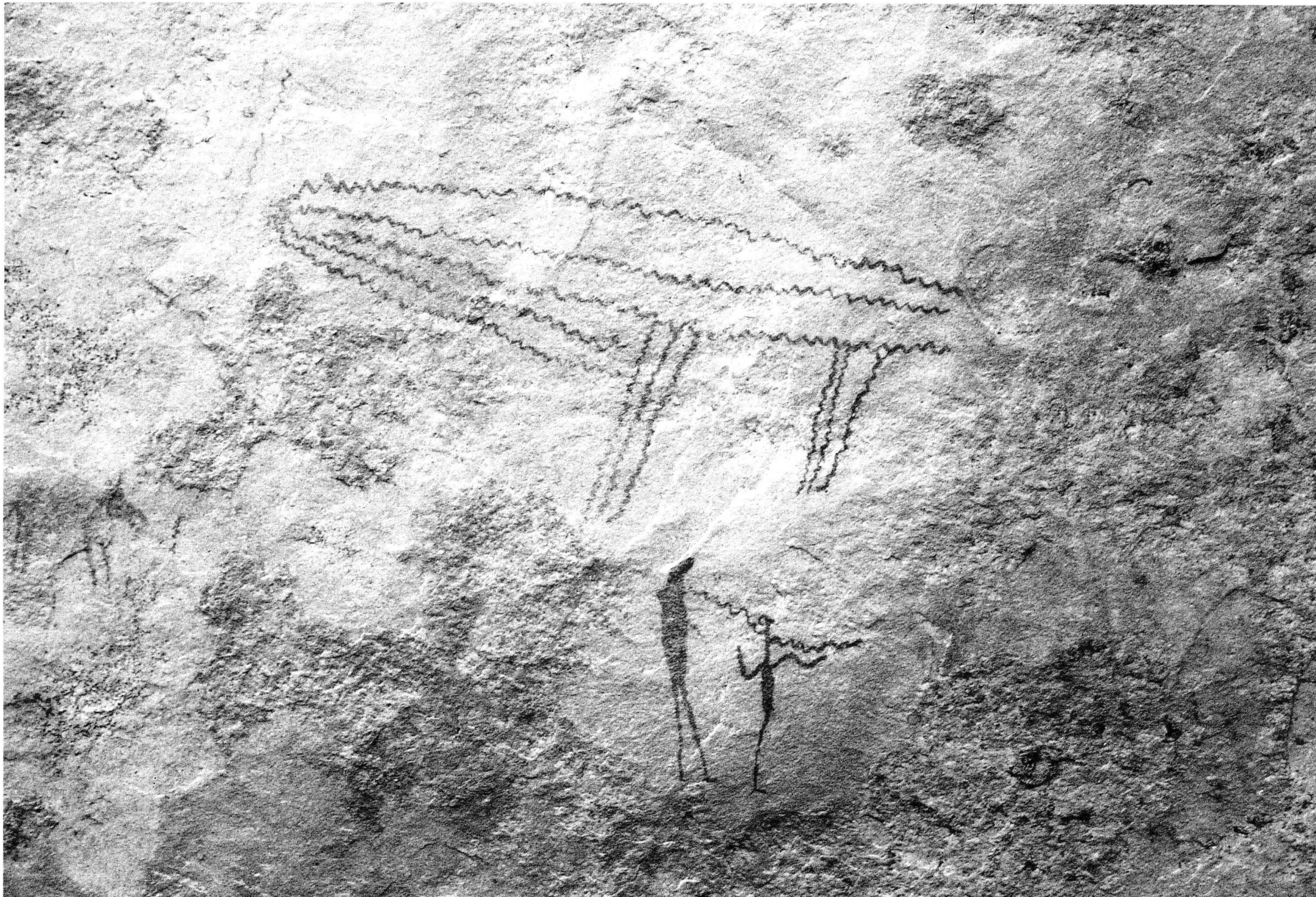

Figure 39: Rain clouds and falling rain. Rain doctors in electrifying conversation? Calvinia.

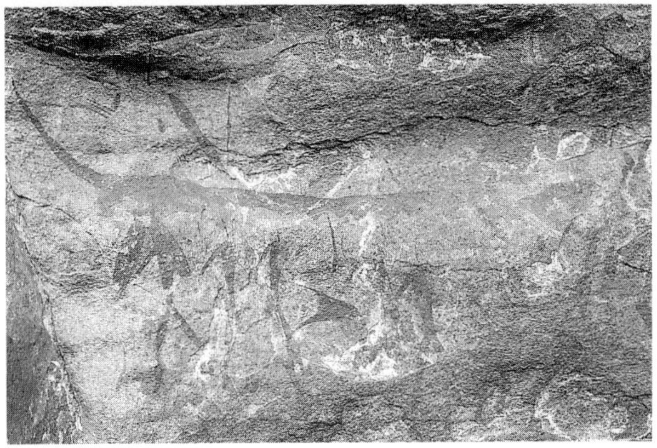

Figure 40:
Spotted rain animal below a rain snake. Transkei.

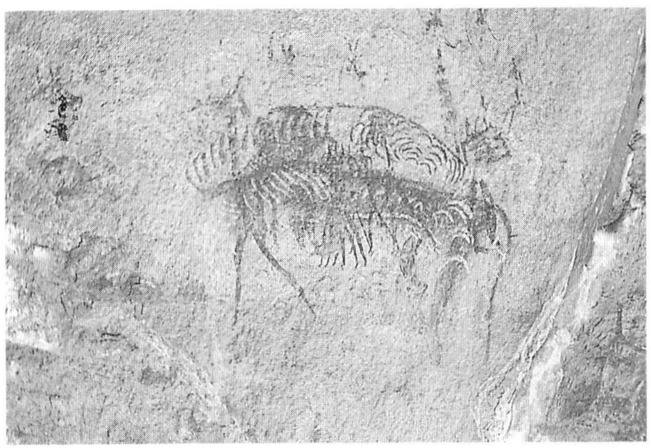

Figure 41: Cloudscape superimposed on walking figures. Bushmen talked about 'ribs of rain'. Ha Khotso, Lesotho.

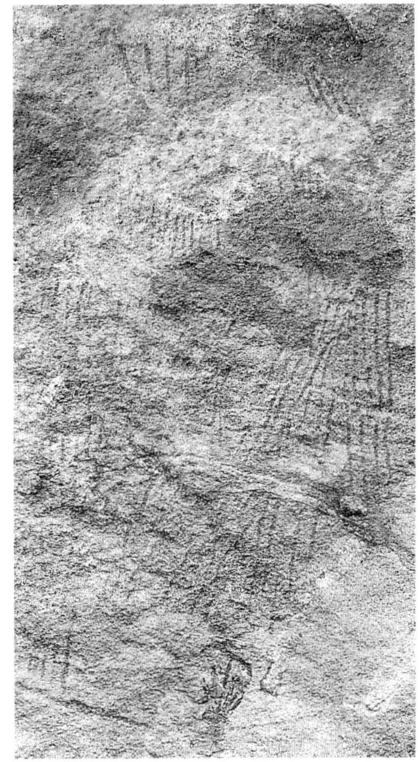

Figure 42: Clouds and rain. Wodehouse.

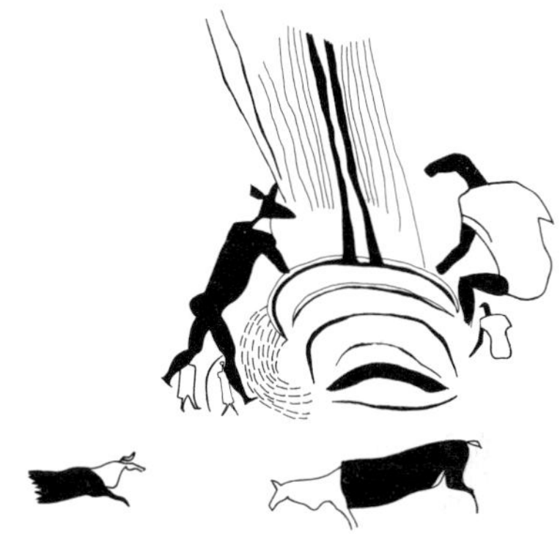

Figure 43: The cyclonic nature of a fierce 'he rain', created or controlled by the therianthropic figures on each side.

however, admit that some might be examples of a more profound 'mythological interpretation'. Lewis-Williams has taken the uncompromising view that all these therianthropic figures represent medicine men or sorcerers in a state of trance (Lewis-Williams, 1981, 93). An alternative interpretation would suggest that at least some may reflect the Bushman belief that 'all animals were once people' (Woodhouse, 1984). There is general agreement that, in certain cases, caps with prominent buck ears are worn (Lee & Woodhouse, 1970, Figure 57; Lewis-Williams, 1981, 77) to assist in either the magical or practical aspects of hunting. Most would agree that the warthog-headed figure driving the herd of eland over the edge of a cliff during a shower of rain probably symbolises a rain doctor or shaman using his magical power to secure this beneficial result.

A combination of clouds, rain showers and two eland-headed figures in a painting at a site in the Wodehouse District was first reproduced as a draw-

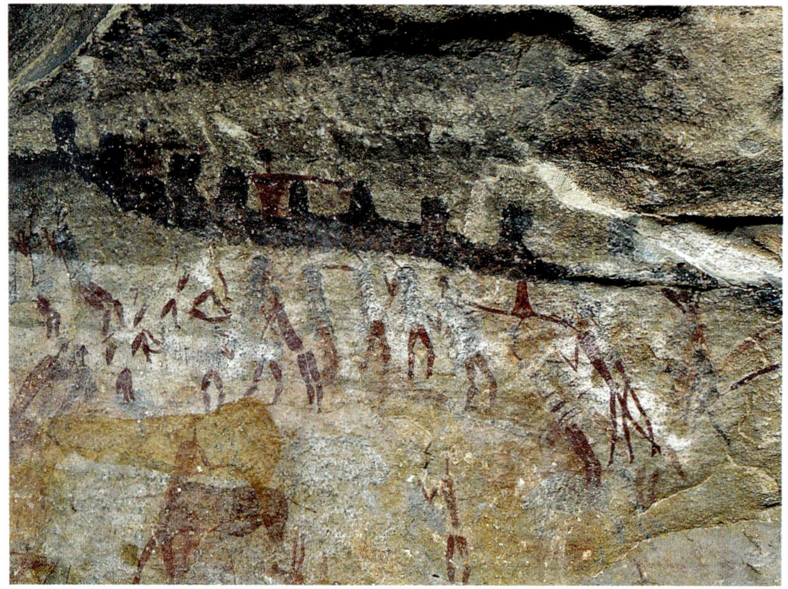

Figure 44:
A line of towering black cumulous cloud on the horizon? Rain dance below. Indwe.

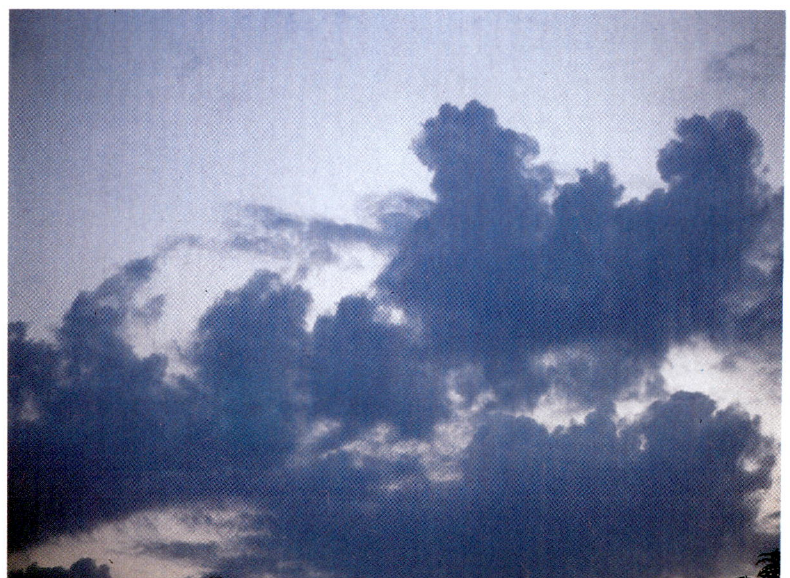

Figure 45:
A possible model for the previous painting.

Figure 46: A Stow copy from Rouxville.

Figure 47: The Rouxville painting copied by Stow. Photographed in 1969.

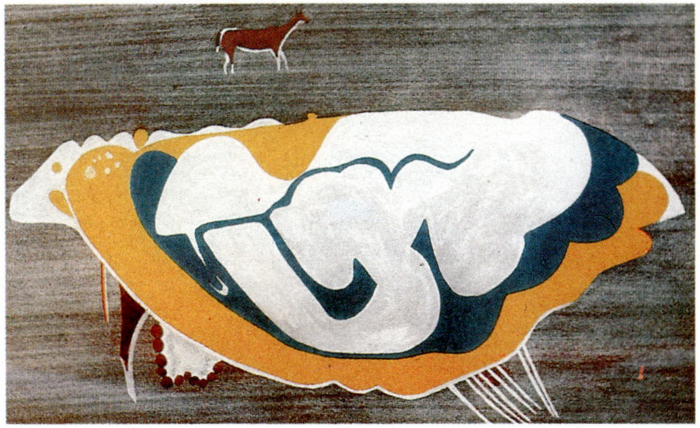

Figure 46

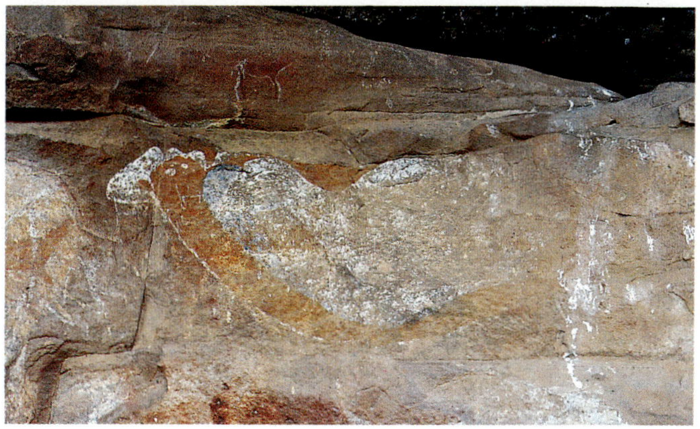

Figure 47

ing in 1970 (Lee & Woodhouse, 1970, Drawing 30) (Figure 43). The two thick vertical lines which are a prominent feature of the painting give the impression of a cyclone — the sort of he-rain dreaded by the Bushmen. This picture is in the same shelter as those shown in figures 36 and 42 so it would seem likely that the shelter was as important a rain-making centre as the one in the Lady Grey district (Figures 25, 29 and 30).

A rare but fascinating feature of the cloudscape of southern Africa is the pendulous form of cumulous cloud known as mammalian cumulous or cumulous mammatus (Figure 12). Such a meteorological phenomenon may have provided the model for a painting in the Wodehouse District and another similar one at the same shelter (Figure 44). The latter example is not quite to specification as the 'breasts' point upwards. It was first interpreted by Helen Tongue (Tongue, 1909) as representing 'barricades defended by soldiers' but it could easily represent a panoply of cumulous clouds on a distant horizon (Figure 45). The busy scene below includes dancers going into trance. There is a somewhat slimmer version of the same subject not far away across the Transkei border (Figure 40). The latter example has a spotted rain animal below it which introduces another variation of the theme of spotted animals to be dealt with in the next chapter. In Lesotho, at Ha Khotso, a huge shelter with a large number of paintings, there is one which appears to depict a 'cloudscape' and falling rain superimposed on a group of walking figures (Figure 41). Stow made a copy of a very intriguing variation on this theme (Stow, 1930, 142) (Figure 46) which has not deteriorated much in the one hundred years between his copy and my photograph (Figure 47). It is interesting that Dorothea Bleek's Bushman informant, when shown this copy, commented: 'The rain with white quartz which has hail' and she went on to write: 'This form of rain animal resembles a tortoise, which was under the special protection of the water.'

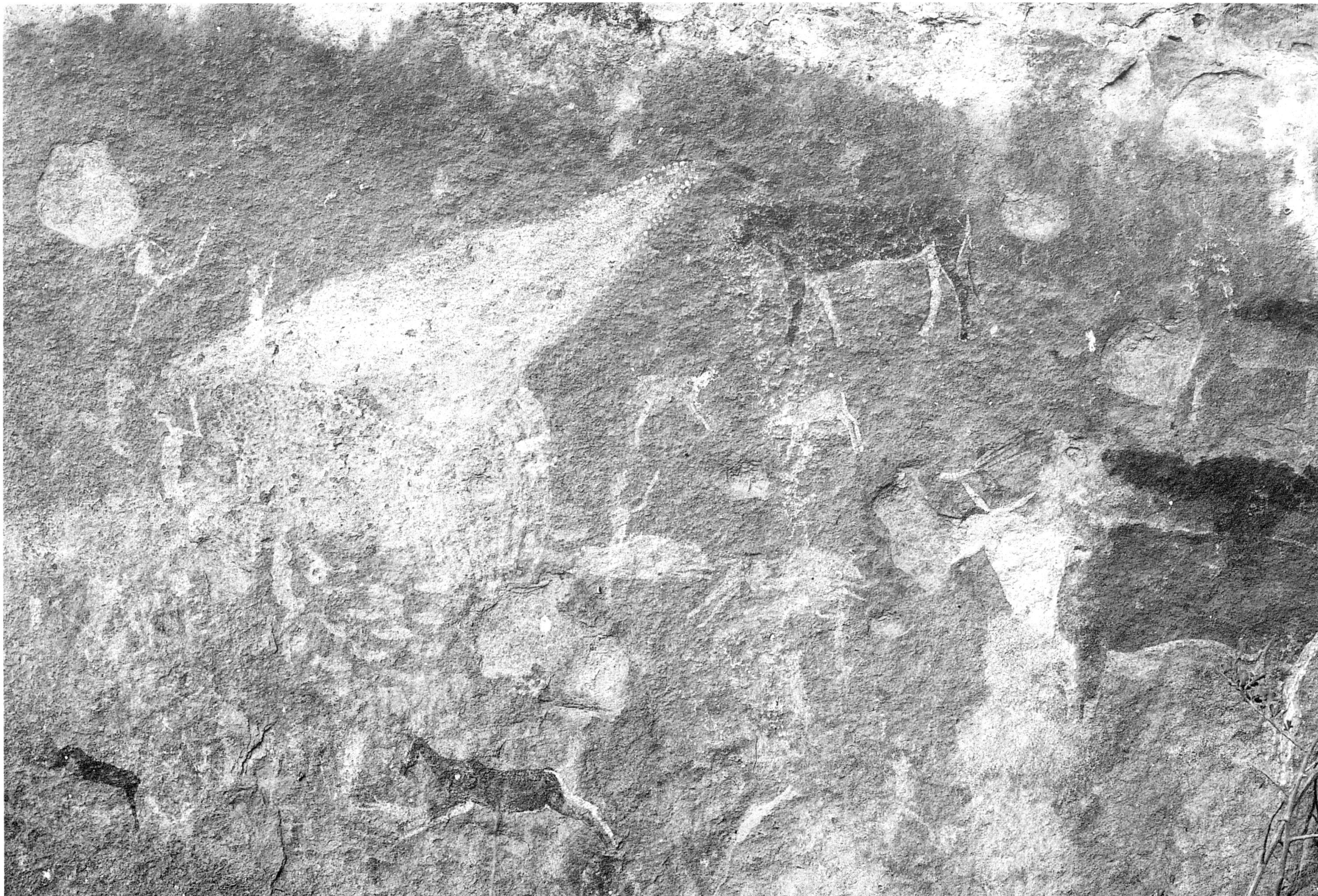

Figure 48: A hailstorm. First published in The Hunter and His Art *by Ione and Jalma Rudner, 1970. Aliwal North.*

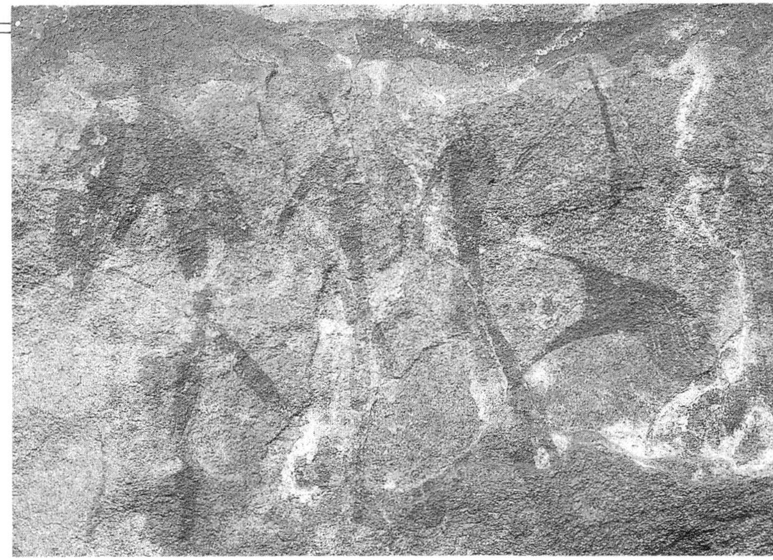

Figure 49

SPOTTED RAIN ANIMALS

In 1958 Jalmar and Ione Rudner traced an interesting and complicated painting in the Aliwal North district (Rudner J and I, 1970, 141). Their tracing is reproduced as plate 54 in *The Hunter and His Art*. They wrote: 'There are some interesting pictures in faded white, showing human figures around a strange white and red dotted shape, which may be a river with a waterfall or a rain cloud — or something entirely different!' In my opinion, and after reviewing all the evidence brought together in this book, I would say that these shapes are not something 'entirely different', but represent a cloud, a sharp shower of rain or even hail, and a dotted rain animal. There are also, appropriately, some white fish (Figure 48), below to the left. Bearing in mind the ferocity of South African hailstorms, it seems appropriate to suggest that the spots represent potential hail not only in this picture but also in the example associated with the 'mammalian snake' described at the end of the last chapter (Figure 44) and in many others still to be considered.

In 1967 Neil Lee and I photographed a scene near Maclear which is also of special interest. It has been reproduced in part and in full on a number of occasions (e.g. Lee & Woodhouse, 1970, dustcover; Woodhouse, 1979, plate 104) to draw attention to various points which it illustrates, such as a man playing a musical bow and a particularly notable man-animal creature with a profile unlike that of a Bushman. There has not, however, previously been any attempt to offer a complete explanation. Now that the existence of spotted rain animals has been identified it is possible to make that attempt. There is the depiction of a rain doctor/shaman represented by the principal white figure that is possibly modelled, as I have suggested elsewhere (Woodhouse, 1974), on a survivor from a shipwreck but with certain animal characteris-

Above: Detail of a spotted rain animal. Transkei. It seems highly likely that the spots represent hail.

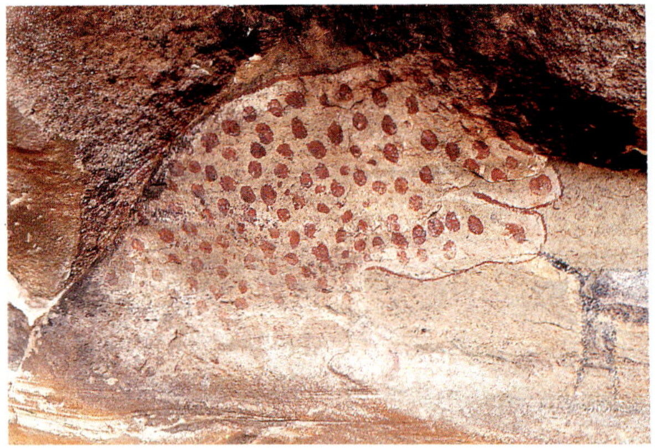

Figure 50: Detail from the rain-making scene with spotted rain animal. Maclear.

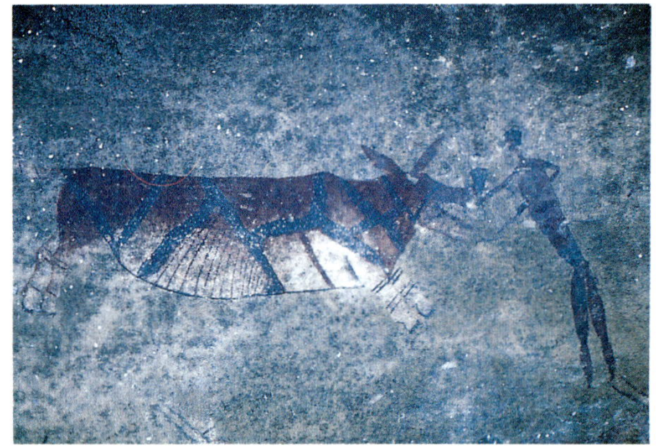

Figure 51: Man equipped with a metal hoe or cutting tool for 'cutting the rain animal', as referred to in Bushman folklore. Aliwal North.

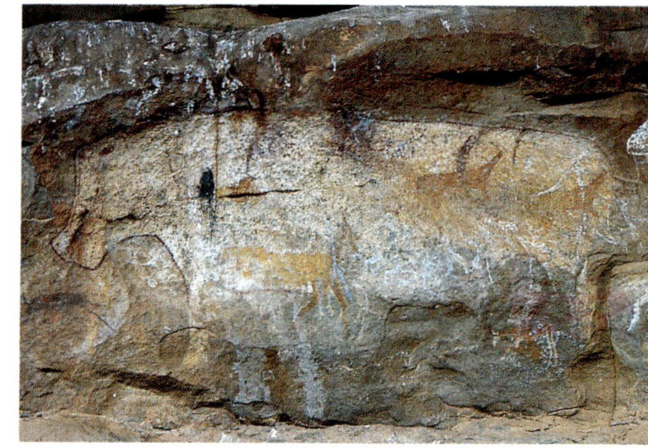

Figure 52: Stow's 'white hippopotamus' — a rain animal. Rouxville.

tics — paws, legs and hooves making him a therianthrope — signifying a trance state. Dancing is taking place to the accompaniment of clapping and the playing of the musical bow.

Professor P R Kirby, authority on indigenous musical instruments, draws attention to the association of the musical bow with rainmaking (Kirby, 1968, 180) and quotes the activities of a youth who summoned up a violent storm emulating the powers of an old rain doctor called *//kunn* who played the bow while lying on his back. Kirby has a photograph of this unusual position (plate 53 ibid) but all rock paintings of musical bows so far recorded have shown the players in seated positions. Dorothea Bleek observed, photographed and described such a position (Bleek, 1928, 121) which differs slightly from the rock painting reproduced here as the resonator in that painting is midway up the bow and is held against the chest instead of resting on the ground. The song sung to the accompaniment of the bow is in the form of a request to the New Moon — named the 'rain-giver' to shake down or even to thunder down water.

The scene also includes the apparent sacrifice of a small buck which is impaled with a long, black stake. Eland and a 'line of magic force' strengthen the composition, and on the left-hand side from a crack in the rock face emerges the spotted rain animal (conceptual) which has been summoned by this activity (Figure 50).

Another painting which has been reproduced in various publications without a full explanation is in Sebaaieni's cave in the Ndedema Gorge made famous by Harald Pager's book of the same name (Pager, 1971, Figure 114) (Figure 54). It is suggested that this is a further example of a spotted rain (or hail) animal. In promising a soft she-rain the rain sorcerer said, '. . . for I will milk a she-rain' (Bleek, 1933, 309). This painting seems relevant to that statement.

In 1958 Walter Battiss published a copy of a savage-looking animal from the Dordrecht district with the caption 'Mythical wild animal approaching fearless Bushman' (Battiss, 1958, 63). A photograph of the same scene was reproduced in 1970 (Lee & Woodhouse, 1970, plate 204) but without a more detailed explanation. It now appears likely that this painting also represents a rain animal. Not only did *//kabbo*, Dr Bleek's informant, talk about milking a rain animal, but also about 'cutting' it. The man in this scene (Figure 57) is appropriately equipped with a suitable artefact and the animal is decorated with bands of spots — hail and a segment of falling rain. The symbolism is irresistible.

There is a large white animal with light rows of red dots in a painting in the Rouxville District (Figure 52) and another painting originally repro-

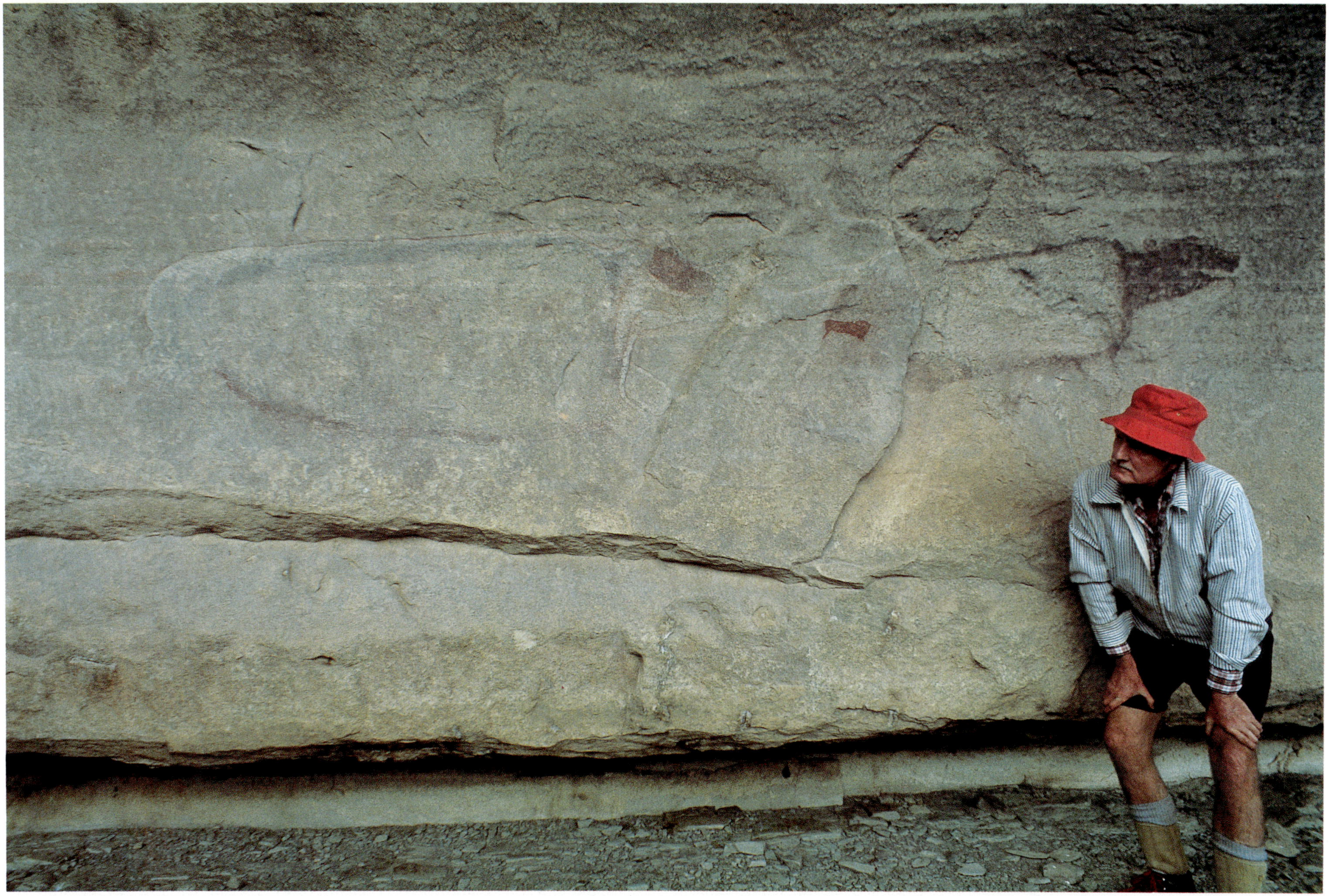

Figure 53: Two large rain animals in a rock shelter in the Wodehouse District. Author provides scale. Right hand animal appears to be emerging from a crack in the rock face.

Figure 54: 'Milking' the rain animal. Ndedema. The original is somewhat more on the diagonal than this photograph.

Figure 55: Two spotted rain animals with humans in attitudes appropriate to the trance dance. Rouxville.

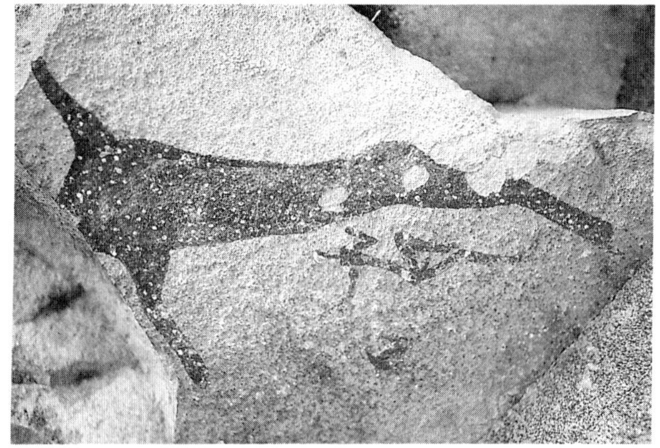

Figure 54

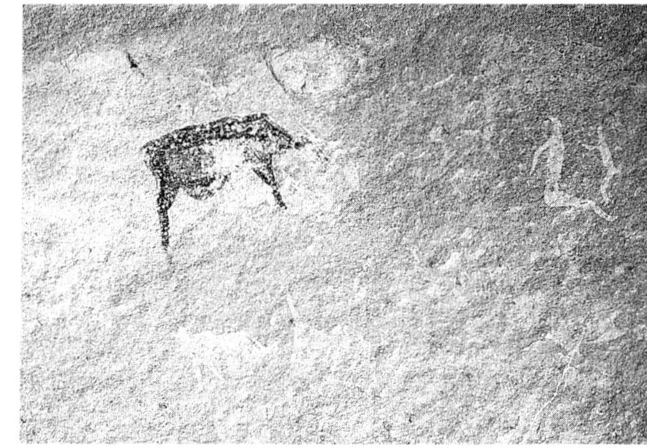

Figure 55

duced by Stow (Stow, 1930, plate 32) in the same district which includes two spotted rain animals and two figures which have appropriate postures for the trance dance (Figure 55).

Alex Willcox reproduced a spotted rain animal from a painting at Giant's Castle (Willcox, 1963, 33) and there is an almost identical animal at Lion's Head, Zimbabwe many kilometres away. Of the one at Giant's Castle he observes: ... perhaps spitting forth rain', which opens up a possible new category of rain animals. But, for the present, it is not proposed to examine all the many paintings of animals, particularly eland, which are painted as bleeding from the mouth or nose, and many of which may have valid 'literal' interpretations as animals that have been hunted (e.g. Lee & Woodhouse, 1970, Drawing 10) or which may have associations with the trance dance.

Perhaps the most remarkable example of a spotted rain animal is to be found in the Transkei (Figure 49). Each red spot is surrounded by smaller white spots and there is a large 'eye' in the head which looks suspiciously like the head of a comet as painted at certain of the sites in the Orange Free State such as that in the Fouriesburg district mentioned in connection with the offering to a rain snake in the third chapter.

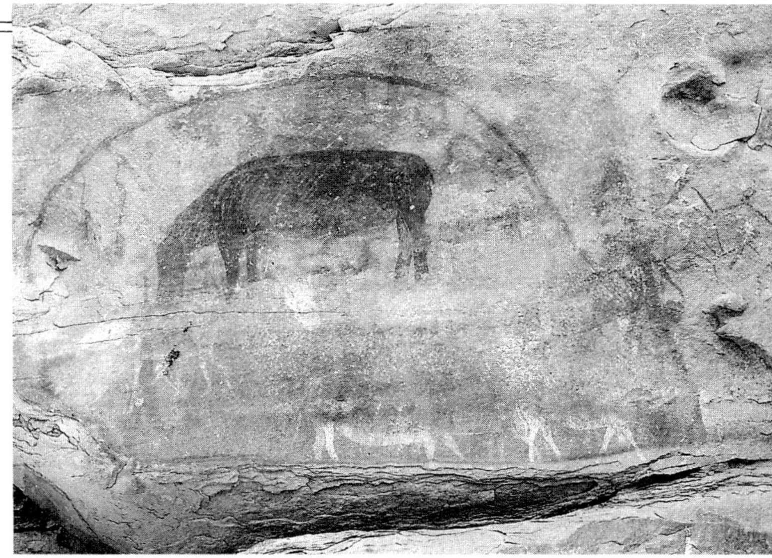

Figure 56

Rainbows

> The Rainbow is yellow in that part that lies above; the piece which seems red lies below. For the Mantis, who is also yellow, lies above, and Kwammang-a lies below.

So recorded Dorothea Bleek in *The Mantis and His Friends* on the basis of the material collected by her father and her aunt (Bleek, 1923, 66). In his book *Rock Paintings of the Drakensberg*, Alex Willcox drew attention to this statement and queried whether the Bushmen had limited colour vision (Willcox, 1956, 55). So far as I am aware the question has not been scientifically investigated, but the range of colours in certain paintings indicates that at least some of the artists could distinguish and employ a very wide range of colours and shades. There is nonetheless an implication in Dorothea Bleek's work that Bushman paintings of rainbows may be represented by as few as two bands. This is the case in the second example used to illustrate the concept of a 'she-rain' in chapter 5 (Figures 20 and 21).

A rainbow with a few more bands, a spotted rain animal and figures with handsome striped karosses, was illustrated in a tracing by Marion Didcott from a colour slide by Neil Lee in *Art on the Rocks* (Lee & Woodhouse, 1970, Drawing 31). At that time Neil Lee suggested that the prone figure drawn in outline lent a mythological element to the painting. Today it would most probably be described as another example of a man in the prone trance position.

There are two other examples of possible rainbows, one featuring a solitary man and the other a solitary animal in the Aliwal North district (Figures 56, 58) but a particularly complicated example is in the Rosendal district of the Orange Free State at the same shelter as the rain snake accom-

Above: Rain animal and rainbow. Aliwal North.

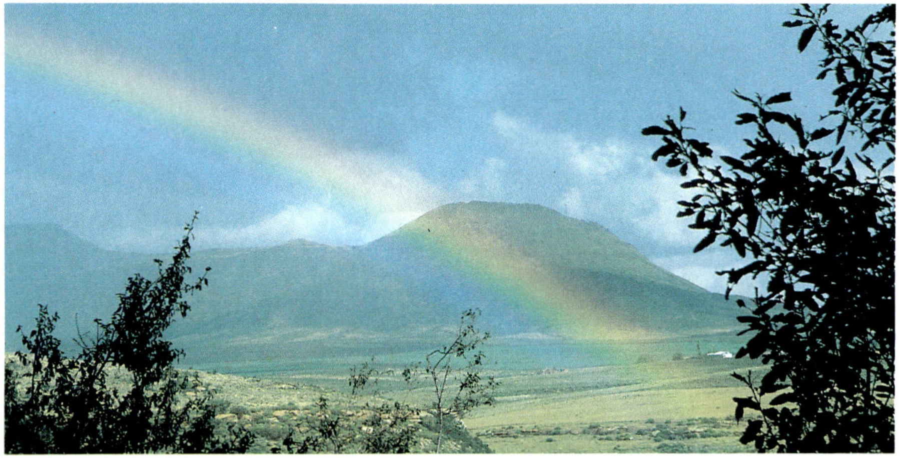

Figure 57

Figure 57: *Rainbow photographed from a painted rock shelter. Wodehouse.*

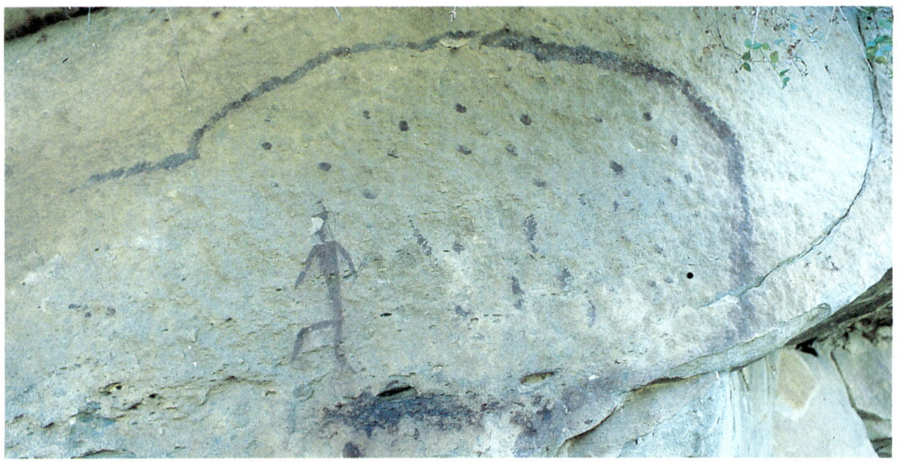

Figure 58

Figure 58: *Rudimentary rainbow? Aliwal North.*

panied by a man with a triangular torso who can be identified with reasonable certainty as a rain doctor or sorcerer (Figure 59). Neil Lee and I first recorded this site in 1983 and the first publication of the painting was as a drawing in the *South African Archaeological Bulletin* (Woodhouse, 1985). The rainbow arches above a painting of straight lines and triangles with an accompanying base of a human figure and lines. These lines have something in common with an almost equally fragmentary painting from the nearby Clarens district described by Mrs G B Rautenbach (Rautenbach, 1967) (Figure 60). To the right of the rainbow is a group of seven dancing therianthropes with, immediately to their right, an upright object of approximately the same height consisting of a succession of Vs placed one above the other, suggesting stylised leaves on a plant. The probable significance of this object or plant form will be dealt with further on, but the therianthropes would appear to symbolise a trance dance associated with the rainbow or a dance from the time when 'all the animals were people' - in this case, clearly rhebok people.

Figure 59: Rainbow with dancing therianthropes. Ficksburg.

Figure 60: Fragmentary rainbow with people. First recorded by enthusiastic amateur archaeologist and palaentologist Georgina Rautenbach. Fouriesburg. Reproduced by kind permission of South African Archaeological Society.

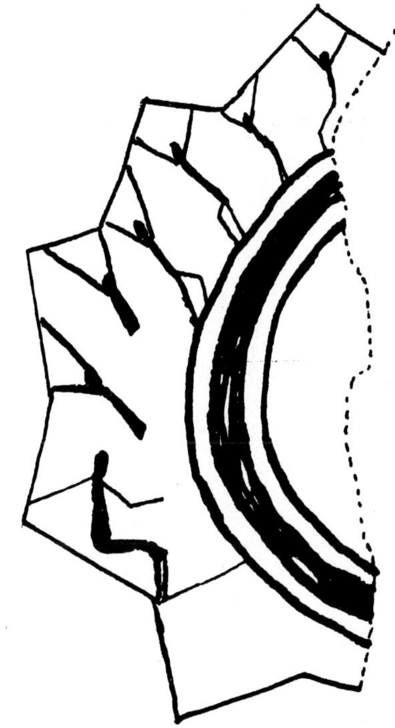

Figure 59

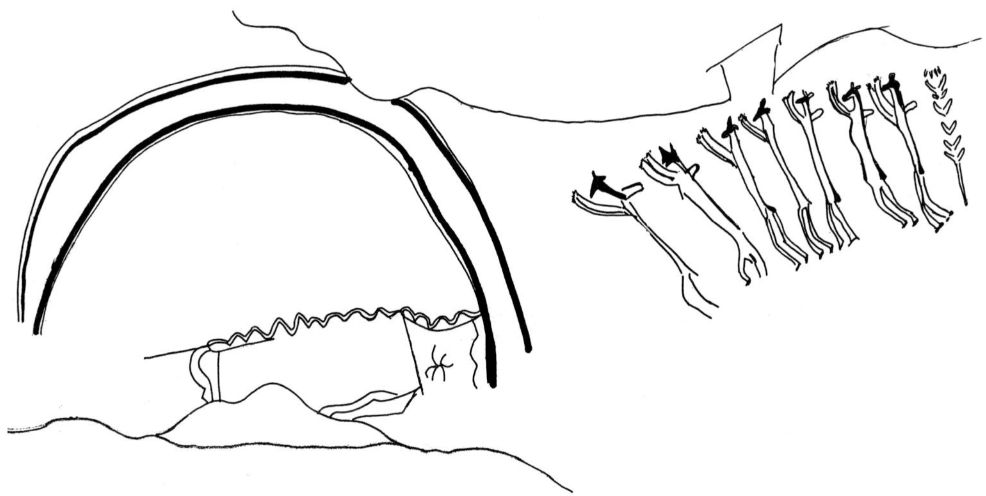

Figure 60

The accompanying photograph of an actual rainbow (Figure 57) was taken from a rock shelter in the Wodehouse District which contains three of the biggest individual animal paintings in the Republic of South Africa (Figure 53). Whether they are rain animals is open to debate, particularly as there is a baby squeezed inbetween them. They resemble hippopotami, but one animal emerges from a crack in the rock face in a rather similar fashion to the rain animal at Maclear. They do not exude any 'showers' but one of the adults and the little one appear to have bleeding wounds. To the right in the same shelter is a group of people, eland and a snake, possibly a rain snake but again without any positive distinguishing features. There is also the rather eroded remains of a spotted animal but it is too indefinite to warrant reproduction. On the whole I incline to the view that this shelter was very much concerned with rainmaking and that the emergence of rain animals from the rock face is a highly significant aspect of Bushman beliefs warranting further investigation.

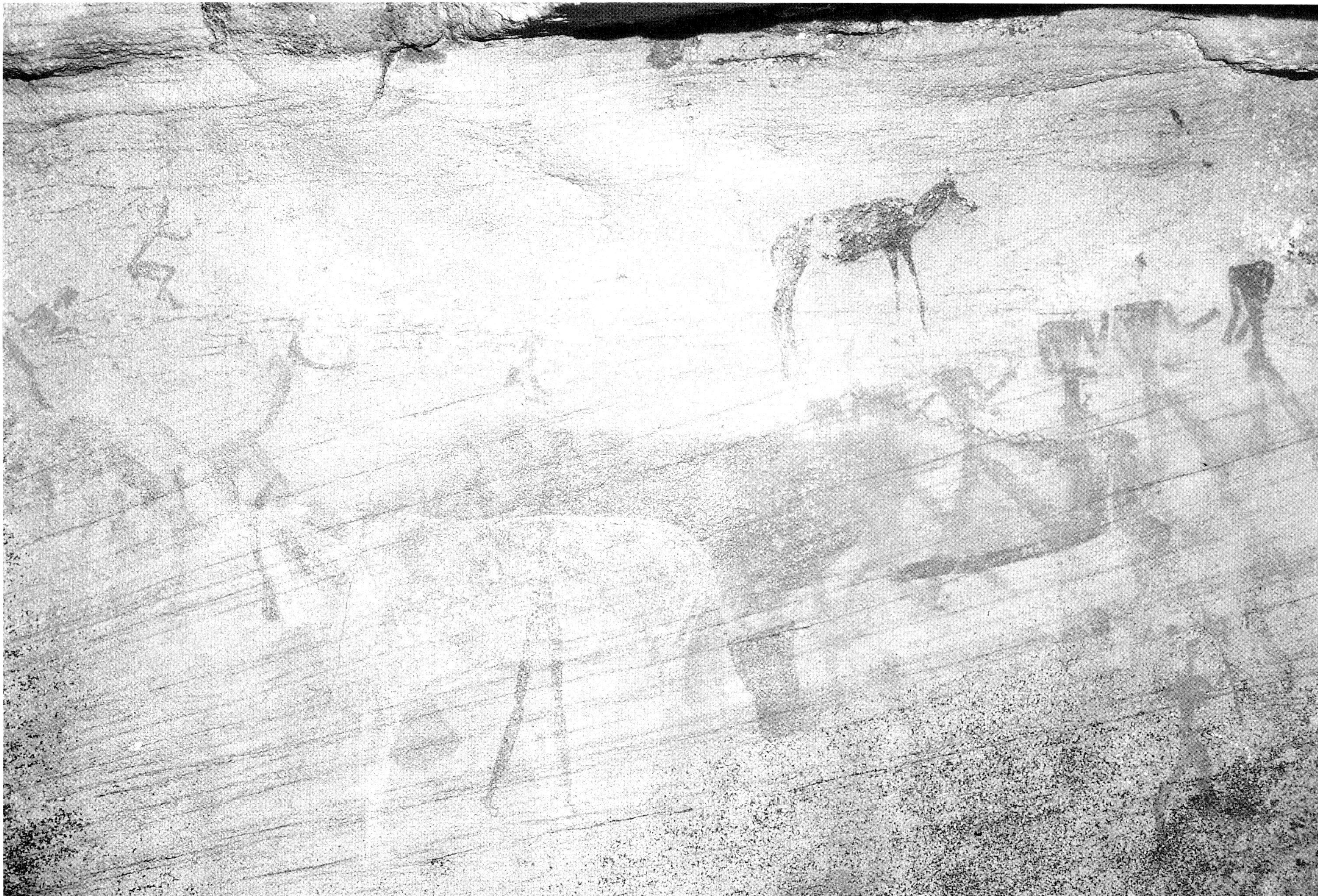

Figure 61: The denticulate pattern, typical of many rain animals, appears to edge the head of a hippopotamus-like creature. Harrismith.

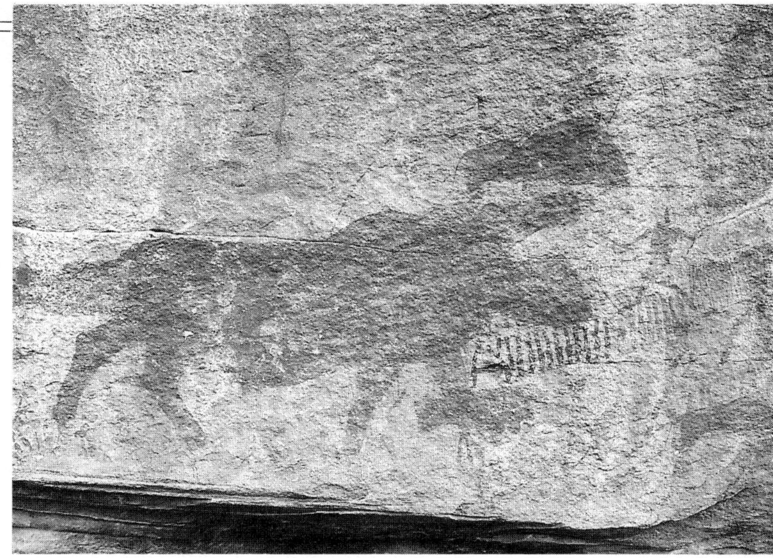

Figure 62

'MASTERS OF THE WATER'

While the hippopotamus-like animals described in the last chapter are not depicted in a context that definitely classifies them as rain animals, they are nonetheless associated with water. They bring to mind the account of the 'Master of the Water' (Stow, 1905, 132) given by a Bushwoman called *Kou'ke* to whom Stow showed a portfolio of his copies of paintings. She applied the term to the creature at Miaputti (Rosenthal, 1953, plate 9) mentioned at the end of chapter 5 and also to the one at Klein Aasvogelkop (Figures 20 and 21) mentioned in chapter 3. *Kou'ke* was the niece of a chief 'Co-ro-ko' and was described by Stow as 'very intelligent' and evidently well-versed in the folklore of her tribe. She said that the 'Master of the Water was an animal of enormous size' that had previously lived among swamps and reeds close to water, and that the Bushmen had set strong fenced traps to capture the animal which, when successful, caused the victim to lash the water into a turbulence creating a spray against which a rainbow was reflected. During attempts to kill the animal, three or four Bushmen were themselves likely to be killed, being crunched between his teeth and then stamped to death (Stow, 1905, 132) (Figure 63).

The term 'Master of the Water' used by *Kou'ke* may be regarded as a synonym for 'rain animal' and her vivid description of the capture and hunting may be symbolic of the dangers attendant upon summoning up such an animal. An actual hunt of a large hippopotamus, or its larger predecessor if there was one, may even have been a task enjoined by a rain sorcerer. There are certainly a number of paintings that qualify for the term. I reproduced three in *Bushman Art of Southern Africa,* one of which is reproduced here (Figure 63) together with four others (Figures 64, 65, 66, 67). One of these

Above: Chosen by artist Ginger Townley Johnson as one of his Major Rock Paintings of Southern Africa (No 27) in his book of that name. Molteno.

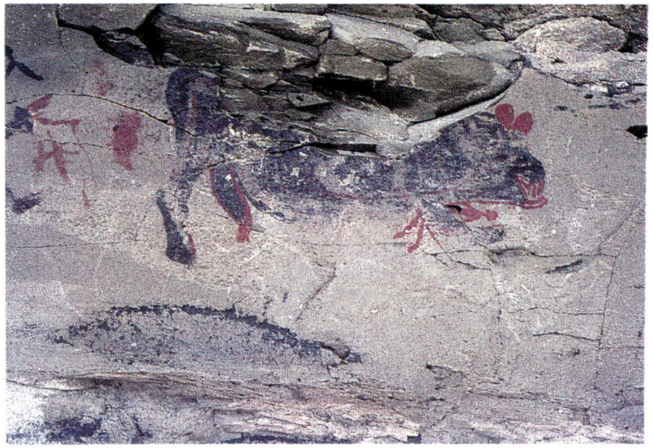

Figure 63: Photographed in 1969. Recorded by Stow in the 1870s. 'The Master of the Water'. Zastron.

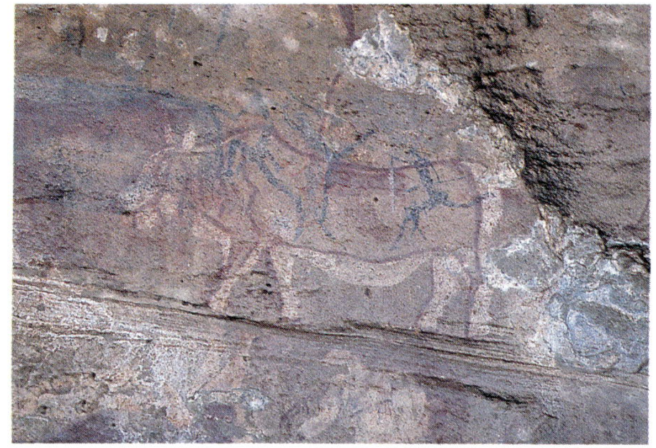

Figure 64: Photographed probably for the first time as recently as 1990. A 'Master of the Water' from the Wodehouse District.

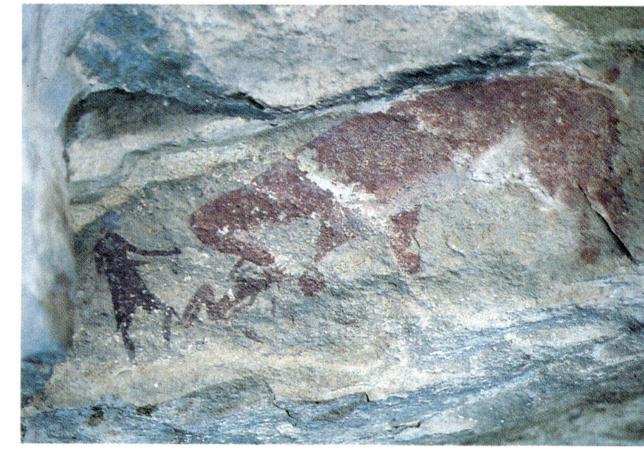

Figure 65: Variation on the theme of the Master of the Water. Aliwal North.

Figure 66: The rain animal is beneath a later painting of a domestic cow. Transkei.

Figure 67: The rain animal, with showers below its neck, that gradually materialises from the confusion of the rock face. Fouriesburg.

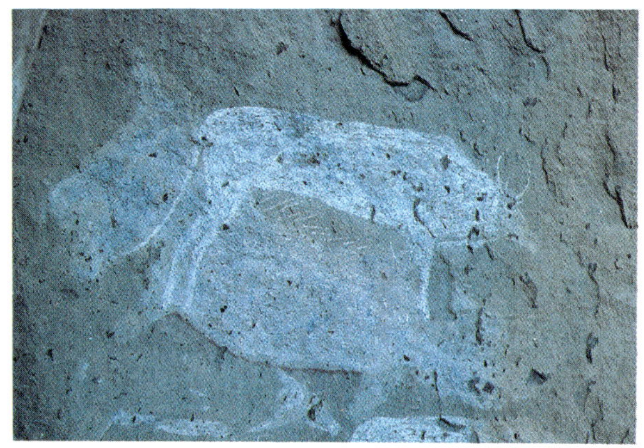

Figure 66

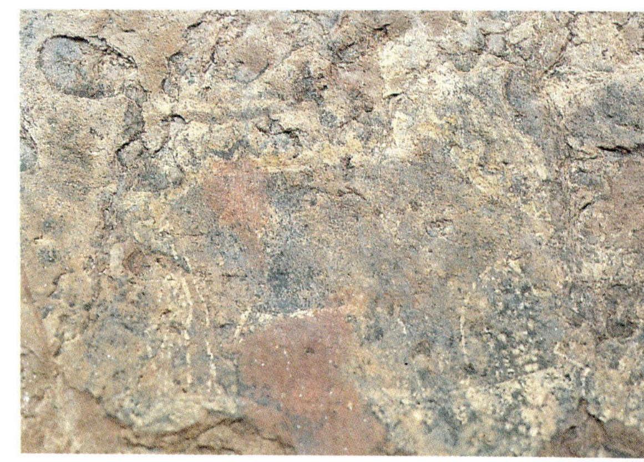

Figure 67

(Figure 65) shows a man prone below the head of the animal. This may represent a man in trance and the animal itself may be symbolic as it may have been seen by the man in trance. The whole scene has an inescapably relaxed air and the upright man appears to be deriving amusement from the predicament of his companion. One of the other examples (Figure 66) has a cow superimposed on it, and is a unique expression of the imagination of the individual artist who created it. The importance of that power of imagination and artistic expression combined with the talent of a single person should never be forgotten when considering this heritage of paintings which have so often been grossly undervalued.

An example only recently recorded in the Warden area is remarkable because it includes along the top of its heavy snout one of the 'wavy lines' that are a predominant feature of the prototype rain animal. It is quite likely that it was painted by the same artist.

The last example (Figure 67) was also recorded recently, having only come to the attention of onlookers after a long, preoccupied stare at what appeared to be a blank rock face.

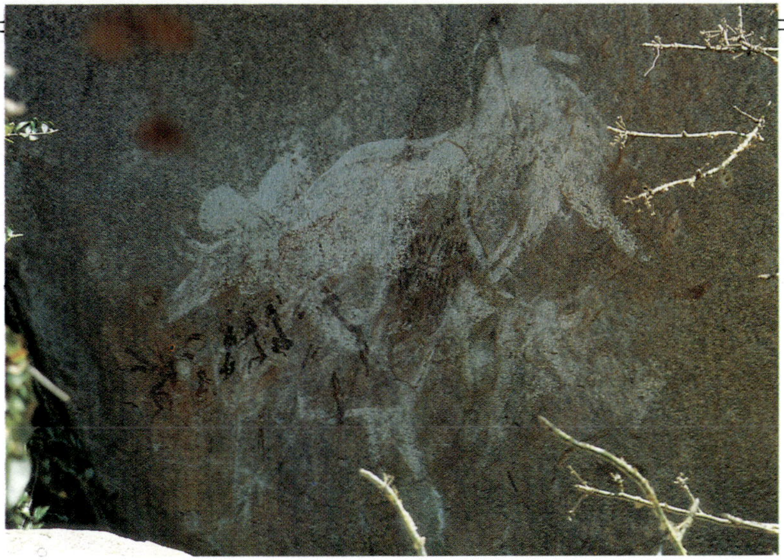

Figure 68

RAIN ELEPHANTS

On a large boulder near the bottom of a flat-topped hill in the Harrismith district of the Orange Free State is a most remarkable painting of an elephant walking on its hind legs (Figure 71). It was first published and described as a 'Cloud elephant with rain pouring from its body' by Erik Holm (Holm, 1961, 166) and subsequently traced and reproduced in black and white by Harald Pager (Pager, 1975, 49). There are no other paintings with this rain elephant but there are many other sites in the vicinity which feature elephants in the more usual paintings. Showers fall from various parts of the anatomy of the elephants just as they do from many of the rain snakes and other animals already examined from other sites.

Erik Holm drew attention to the similar concept expressed in this painting and others in North Africa and Namibia and even to the mobiliary art on a fragment of ivory from the archaeological site at La Madeleine in the Dordogne. These may seem somewhat far-flung examples to have any direct relationship but, fortunately, there are others that are nearer. The painting that most resembles the one at Harrismith is at Mrewa cave between Harare and Mtoko in Zimbabwe, a spectacular site with many interesting paintings which include the rain elephant with 'showers' springing from its tail and belly in marked similarity to the Harrismith example. This elephant is part of a frieze which includes a number of wild animals such as bushpig, rhinoceros, buffalo and kudu as well as people, but facing the elephant is another rain animal which seems to be modelled on a roan or sable antelope. There is a man stooping in a 'trance dance' attitude in front of this animal (Huffman, 1983). Behind the elephant is a fat figure — which is typical of a number of Zimbabwean paintings — and about which that shrewd observer, Mrs Elizabeth Goodall (née Mannsfeld) wrote: 'The legs are always strad-

Above: Rain elephant. Saffron Walden, Harare.

Figure 69: Rain elephant. Mrewa.

Figure 70: Rain elephant. Giant's Castle.

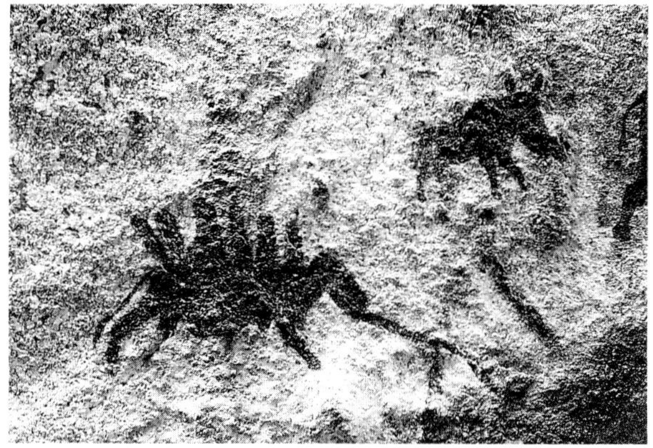

Figure 69

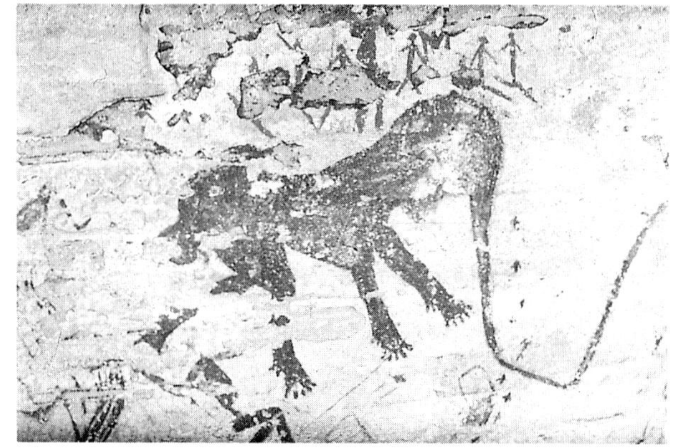

Figure 70

dled, and often two or more streams or bands emanate from the lower part of the body' (Goodall in Summers, 1959, 93).

There would seem to be a relationship between the concept represented by these paintings and the notion of 'river source' discussed in chapter 7. It will be considered in more detail in a later chapter. Another of these fat figures is associated with two very strange elephants (As in Figure 69) in the left-hand corner of the same frieze. This fat figure has breasts so it is reasonable to assume that both figures are female and intended to be pregnant. But at other sites there are similar figures that are undoubtedly male. A further painting of the 'strange elephant' located not far away from the others but — although clearer — does not help to elucidate the meaning (Figure 69). Mrs Goodall classified it as 'mythological' and left it at that but there seems little doubt that it is part of the rain animal concept, particularly as unnaturally long trunks are a feature of two of the animals. In the case of the third, the trunk is cut off by exfoliation of a rock fragment. The vertical additions to the backs of these elephants continue to defy explanation, although they could represent clouds like those on snakes in the Indwe District. All the same, the long trunks seem to accord with the river-source theory.

There are elephants with similar unnaturally long trunks at several scattered sites. One is at Barnes Shelter in the Giant's Castle Nature Reserve, Natal (Figure 70), first reproduced by Alex Willcox (Willcox, 1963, 33) where he illustrated it under the sub-heading 'Mythical creatures' and commented: 'Elephant-like figure but with claws and impossible trunk.' A second example is in the Wepener district of the Orange Free State and a third is near Aliwal North where it appears to be accompanied by another rain animal which is heavily over-painted. In the same region, tucked away under a ledge, is a further example painted in yellow which does not have a long trunk. In fact its rather stout trunk is partly doubled back, but it has other characteristics of rain animals in the form of white spots and white lines (Figure 72) while there are traces of a 'shower' below the front of its belly and a white streak extending from its trunk.

When in the Cape Province, a visit to the west, in particular to the Piketberg district, will lead to one of the finest friezes of rain elephants, in three separate groups (Figures 73, 75). The existence of these paintings was first reported in the *South African Archaeological Bulletin* in September 1961 when an excellent copy of part of the scene by R Townley Johnson was reproduced on the cover. Professor Goodwin commented that the 'network' surrounding the elephants reminded him of a scene on Fenga Hill in the Kondia District of Tanzania and that it was 'possibly the artist's conception of elephants caught in a group of pitfall traps'. A copy of that scene has been

Figure 71: Rain elephant. Harrismith.

Figure 72: Rain — and hail — elephant. Aliwal North.

Figure 73: Rain elephants. Porterville.

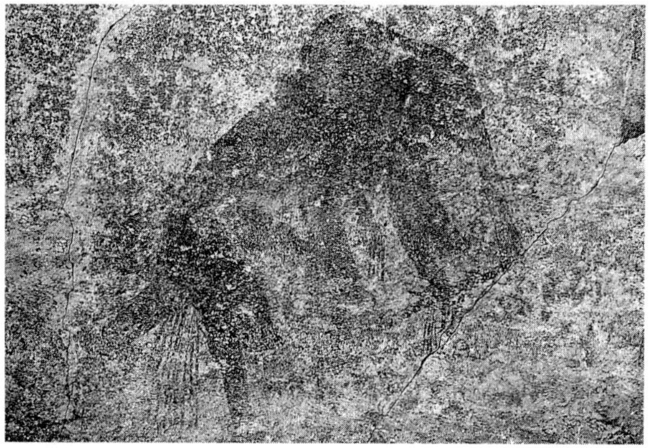

Figure 71

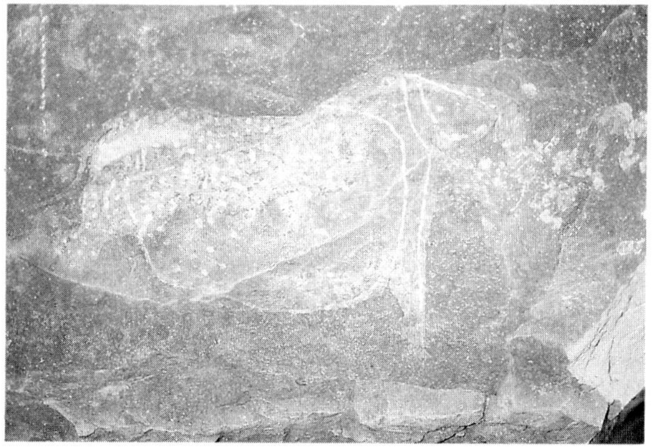

Figure 72

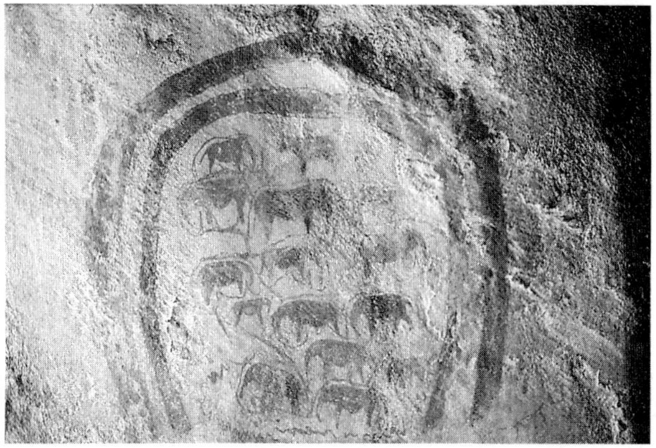

Figure 73

published (Fosbrooke *et al*, nd) and it could equally well be explained as representing two rain elephants. In 1979 I still perceived them as 'man-made enclosures' (Woodhouse, 1979, 110). In contrast, Townley Johnson hinted at their 'symbolism' (Johnson, 1979, 68) and it was an article, 'Elephants in Boxes' (Maggs & Sealy, 1983, 44) based on a paper delivered at a meeting of the Association of Archaeologists which provoked me into thinking more deeply about these paintings and publishing a photograph of one of the groups with the caption: 'Rain elephants. The elephants in the sky that send the rain — with rainbow' (Woodhouse, 1984, 38). The Maggs & Sealy article advanced the view that the patterns round the elephants were a representation of the 'pulsating geometric forms' seen by people in trance — a suggestion that has been further developed since Lewis-Williams and Dowson (1988) but which is not incompatible with their interpretation of water as a universal symbol similar to the views of Professor Goodwin.

There is a 'rain elephant' painted in a similar manner to the 'river source' rain animals discussed previously. It is in Keurkook Kloof in the Worcester district and was first published by Bill Trew in 1984 (Trew, 1984, 133). He kindly guided me there in 1968 when my photograph of the animal was taken. Although there are many other paintings of elephants in the Cape, this, and the ones discussed above are the only ones that I would positively

Figure 74: Elephant with vultures on the 'ceiling' of a painted rock shelter. Uitenhage.

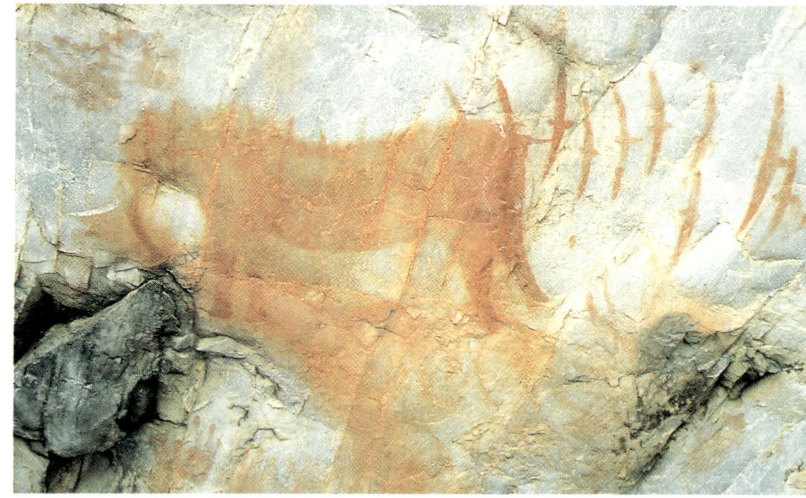

Figure 74

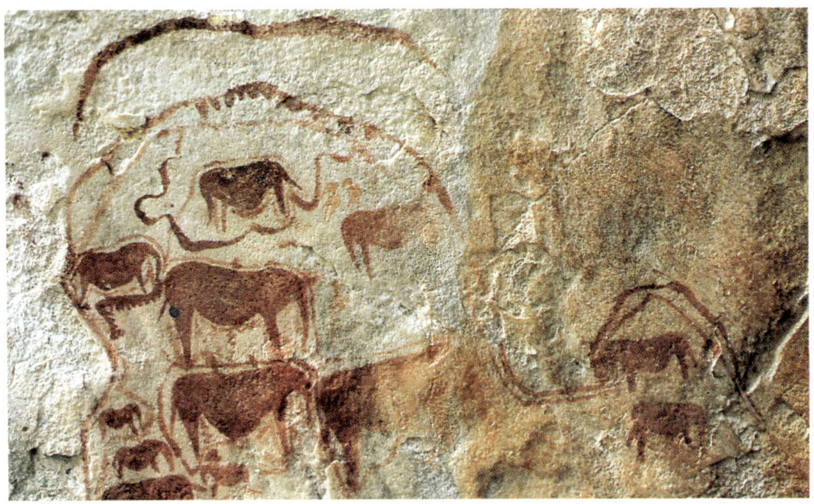

Figure 75: Rain elephants. Porterville.

Figure 75

identify at present as 'rain elephants'. There is, however, one near Uitenhage that must come very close to qualifying (Figures 74 and 78). It was first recorded by enthusiast Ludwig Abel of Port Elizabeth and published by Alex Willcox (Willcox, 1963, plate 13). The elephant is painted on the 'ceiling' of the shelter and is accompanied by a flock of flying birds which Willcox suggested might be vultures, an opinion with which I agreed when I published a picture of them in 1984 (Woodhouse, 1984, 103). There is no feature of the elephant that enables it to be positively classified as a rain elephant, except that the position in which it is painted, together with the wheeling flock of vultures, gives it every appearance of being high in the sky — like a huge rain cloud.

This cloud feature is repeated in a painting not far from Harare among large granite boulders at Somerby where what appears, at first glance, to be a straightforward representation of an elephant in outline is seen, at second glance, to have an additional outline above and partially superimposed upon it. This by itself might not be very convincing, but further along the rock face are hippopotami and what Mrs Goodall has called 'eel-like fishes' together with people, buffalo and kudu. There are also kudu superimposed on the elephant but I doubt whether there is a relationship between them although, by coincidence or intention, there is a similar small buck painted

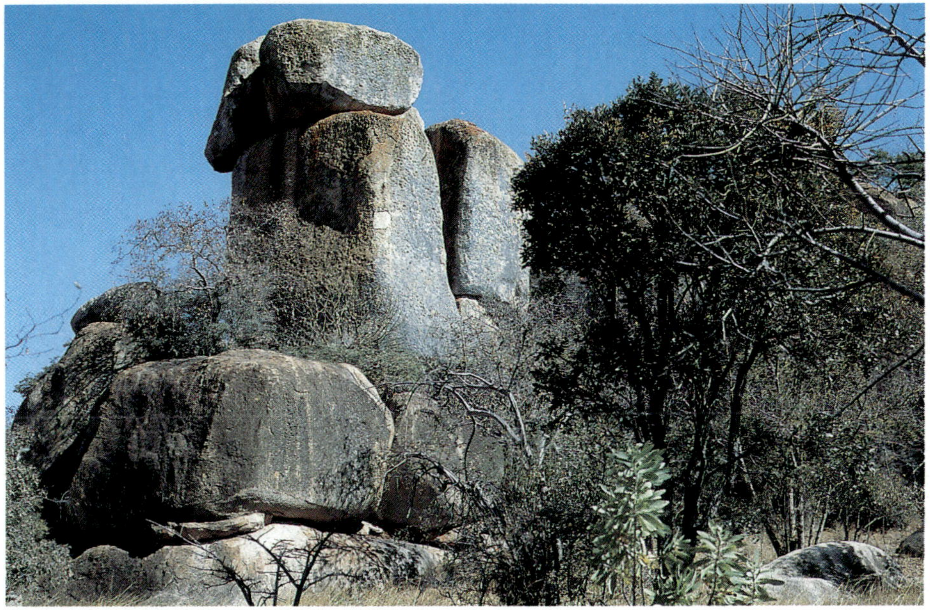

Figure 76: The landscape. Saffron Walden, Harare.

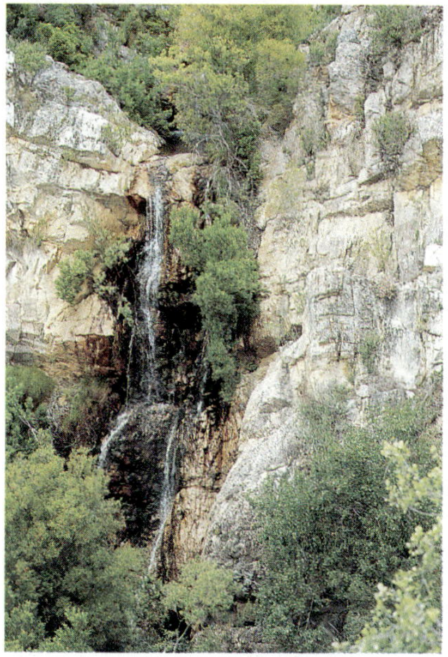

Figure 77: Appropriately adjacent to the site of the rain elephants. Porterville, Piketberg District.

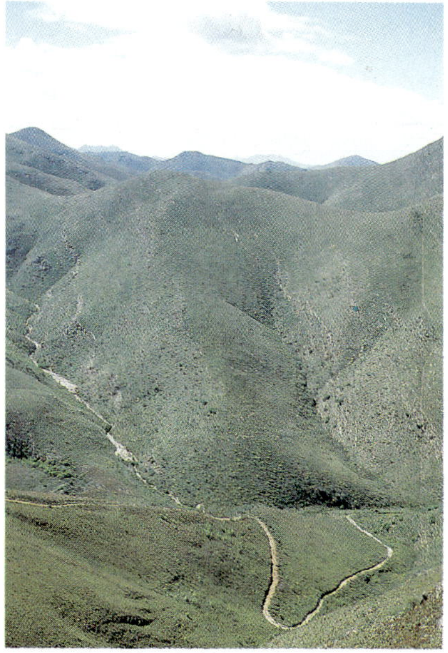

Figure 78: The environment of the painting in Uitenhage depicting elephant with vultures.

in red monochrome in the centre of the large white elephant at Philipp Cave in Namibia quoted and reproduced by Erik Holm as an example of the 'cloud elephants'. A somewhat similar painting within a few kilometres of the Somerby site (Figure 68) appears to have a shower of rain falling from it as does at least one of a procession of large elephants in the Mtoko district which was first reproduced in 1970 (Lee & Woodhouse, 1970, 27).

At various sites in Zimbabwe there are other paintings of elephants that may be interpreted as rain elephants because of their peculiar features or associations. At Kulubahwa (formerly Majenje) is an elephant divided into segments like the rain animal in the Aliwal North district. At the well-known site Bushman Point, Lake McIlwaine, Harare is an elephant associated with fish. In the Chibi area near Nyanda (formerly Fort Victoria) are red elephants outlined in white which gives them a cloud-like appearance and in the Gwanda district there is a representation of a herd of elephants with unusual markings on their heads and trunks which may symbolise rain. At Pomongwe in the Matopos near Bulawayo, Cran Cooke reported an outline elephant similar to that at Somerby (Cooke, 1963). It was reproduced on the cover of the *South African Archaeological Bulletin* in November 1963.

Just as there was a close contact period of several hundred years between Bushmen and Bantu-speakers in parts of South Africa, there must have been a similar situation in Zimbabwe. The English archaeologist M C Burkitt in his classic work *South Africa's Past in Stone and Paint* describes his visit to the well-known site Domboshawa near Harare and records the local tradition of rainmaking and the offerings that were brought there by supplicants who waited until a sign appeared that their offerings had been accepted and that rain would be forthcoming. The sign took the form of smoke rising from a fissure at the top of the huge granite dome in which the rock shelter is situated. This could only be accomplished when there was an easterly wind — which normally brought rain.

While Burkitt was at Domboshawa, Professor (then Mr) Goodwin discovered a painting nearby which depicts a rain-making ceremony involving a tree. It was subsequently copied by Mrs Goodall and reproduced as Figure 30 in *Prehistoric Rock Art of Rhodesia and Nyasaland*.

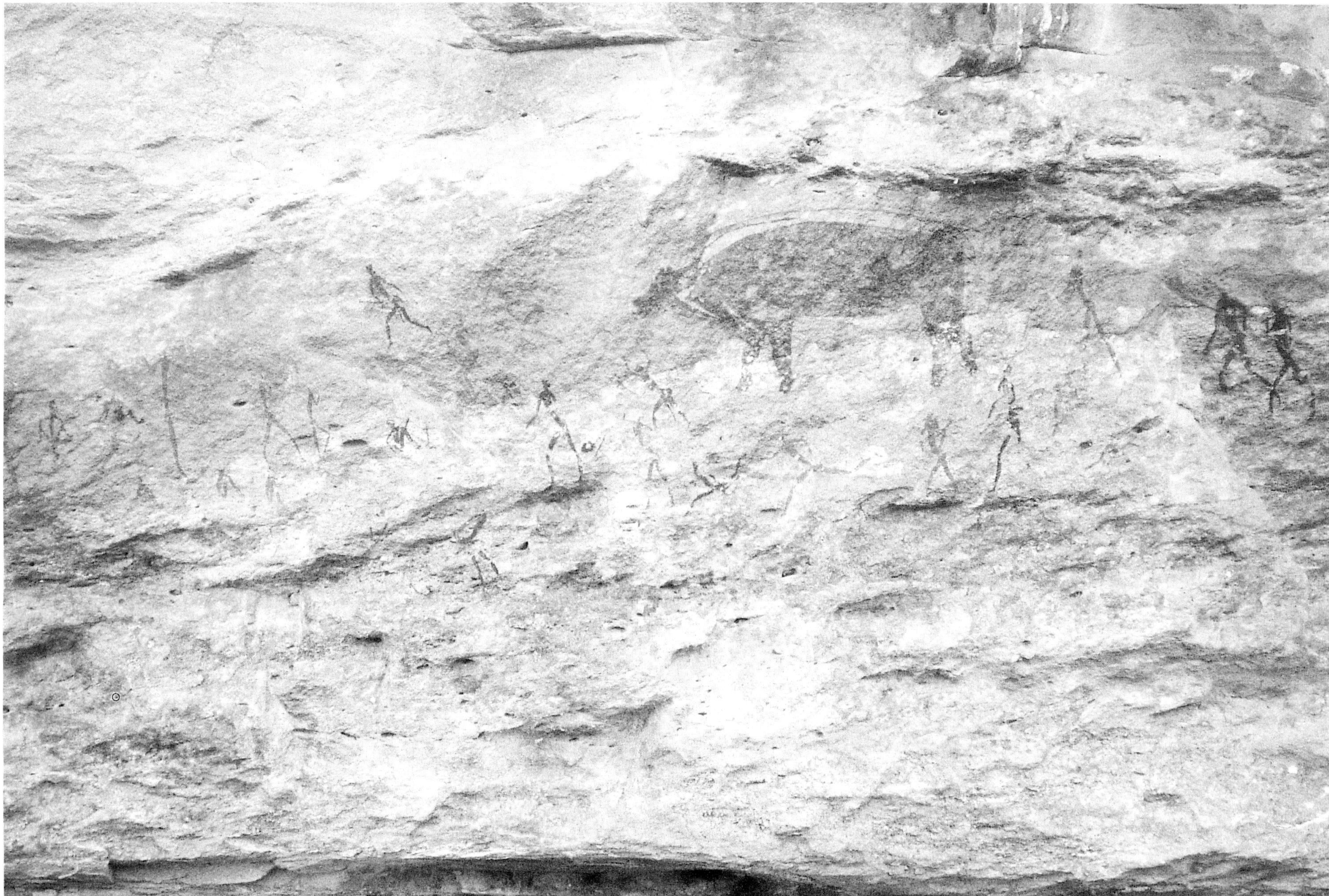

Figure 79: Leading/Driving the rain animal. Harrismith.

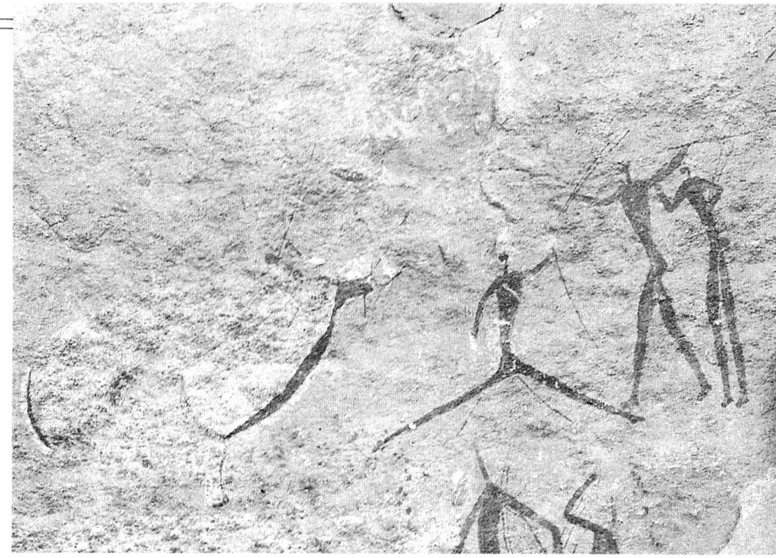

Figure 80

LEADING THE RAIN ANIMAL

The 1870s were fruitful years for the study of rock paintings. G W Stow and J M G Orpen were making copies and Dr W H I Bleek and his sister-in-law, Lucy Lloyd, were studying the Bushman language and recording the folklore. One painting in particular (Figure 81) copied by magistrate J M G Orpen in Lesotho was shown to different informants. The first was Qing, son of a Bushman chief who had managed to escape the extermination of the rest of his tribe and who entered Orpen's employment. He explained the painting by saying: 'That *animal* which the men are catching is a *snake (!)*. They are holding out charms to it, and catching it with a long riem. They are all under water, and those strokes are things growing under water'. He then went on to say that the people had been 'spoilt' by a dance (Orpen, 1874, 10). Dr Lewis-Williams has suggested that 'under the water' is a metaphor for 'in trance' and that the dance was the well-known 'trance dance' by which this state was attained (Lewis-Williams, 1977b).

The second group of informants to whom the copy was shown were Bushmen from the Calvinia district staying with Dr Bleek. Their explanations as recorded by Dr Bleek were:

> The paintings from the cave Mangolong represent rainmaking. We see here a water thing or water cow . . . They then charm the animal and attach a rope to its nose — and in the upper part of the picture it is shown as led by the Bushmen, who desire to lead it over as large a tract of country as possible — their superstition being that wherever this animal goes rain will fall. The strokes indicate rain. Of the Bushmen who drag the water cow two are men (sorcerers), of whom the chief one is nearest to the animal.

Above: Leading/Driving the rain animal. Harrismith.

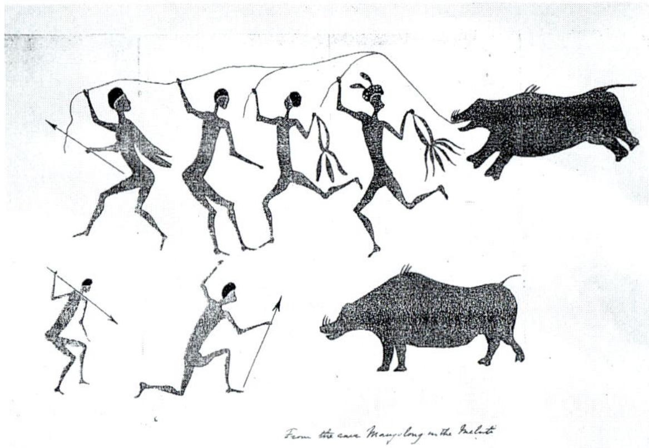

Figure 81: Enticing the rain animal and leading it over the land needing water. Copy made by magistrate J M G Orpen in Lesotho in 1873 and explained by Bushman Qing.

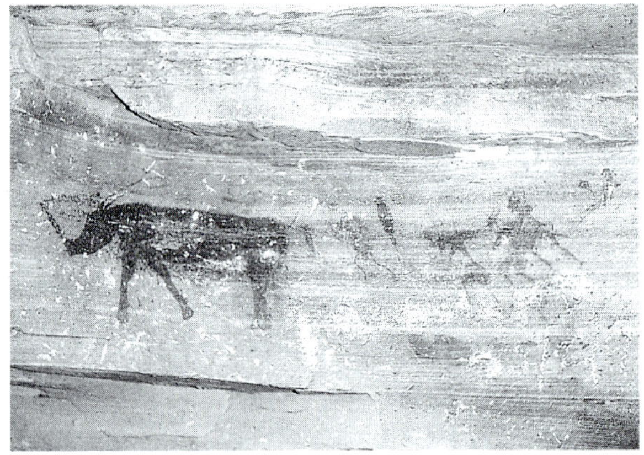

Figure 82: Leading/Driving the rain animal. Wepener.

Figure 83:
Bottom and bottom opposite: Rain animal being led out to generate a Drakensberg storm and so to obscure the spoor of horses and cattle after a successful Bushman raid on frontier farmers. Formerly Underberg, now in the Natal Museum, Pietermaritzburg, to whom grateful acknowledgements are made for permission to photograph in 1967.

In view of the large number of 'rain snakes' recorded earlier it is of interest that Qing used the word 'snake', obviously to Orpen's surprise, when shown the painting. The strokes which were present in Orpen's original copy, but which were not reproduced in the printed version, conveyed the impression of water or rain to all informants, and the concept of the rain animal being 'charmed' and led across the countryside was commonly accepted. There was a second similar animal in the painting which was being driven rather than led — presumably an equally efficacious procedure.

A classic example of the concept of a charmed rain animal accompanied by a graphic depiction of an actual event was published by Patricia Vinnicombe in her book *People of the Eland* (Vinnicombe, 1976). She illustrated it with an old black-and-white photograph taken while the painting was still *in situ*. The whole scene was subsequently cut out of the rock face, split into two halves and moved to the Natal Museum, Pietermaritzburg (Figure

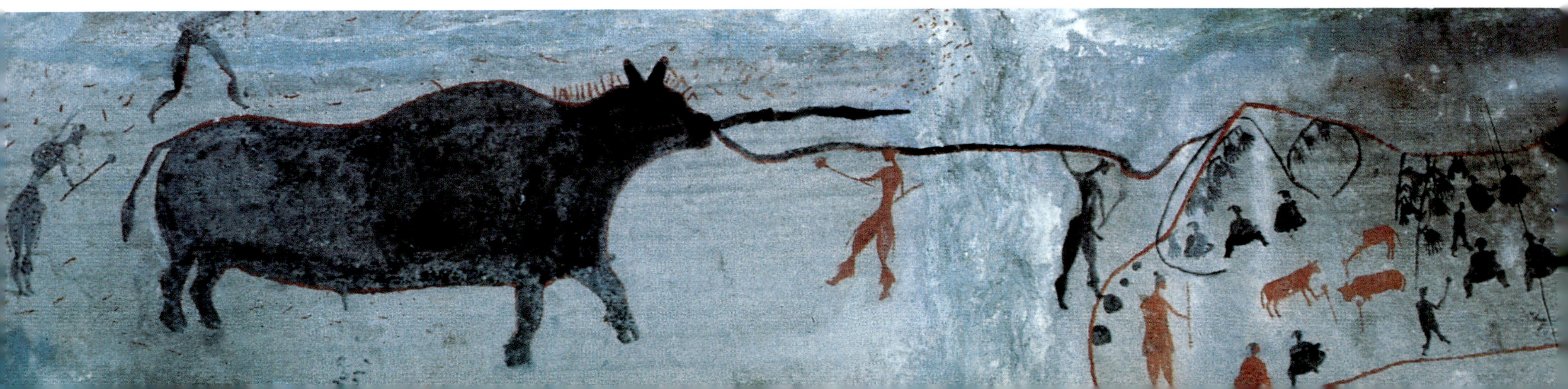

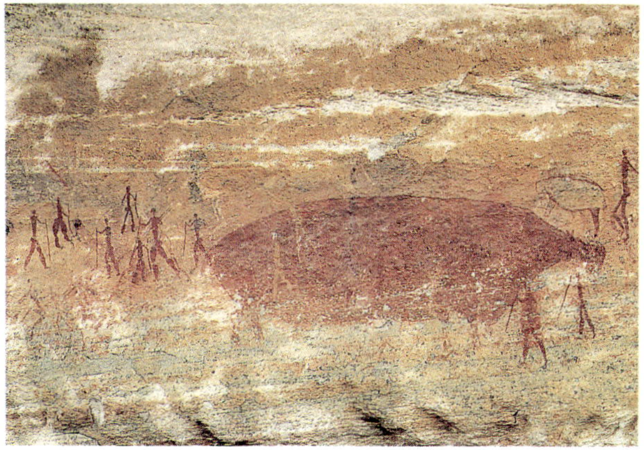

Figure 84: Leading/Driving the rain animal. Lady Grey.

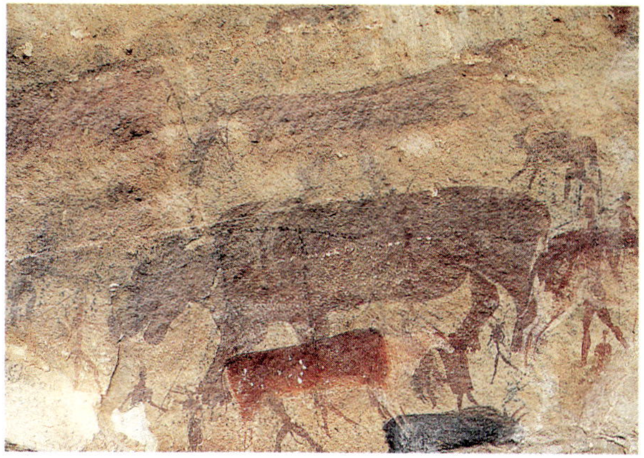

Figure 85: Leading/Driving the rain animal. Aliwal North.

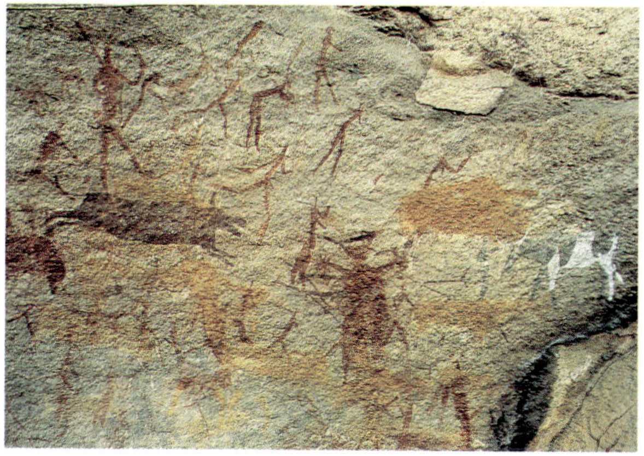

Figure 86: Leading/Driving the rain animal. Transkei.

83). As Vinnicombe wrote, the scene illustrates Bushmen in a rock shelter with their carrying bags and other possessions. On the right, a successful mounted raiding party returns with cattle and horses. The leader can be distinguished by a fine leopard-skin cloak and his white horse — a leaf out of the book of British military officers. To the left a rain animal is led by a long riem or thong symbolising a heavy rain shower.

During the last century it was customary for cattle and horse-stealing to be timed to coincide with one of the frequent Drakensberg storms which would obliterate the spoor of the raiders and their booty.

There are many other examples of leading or, in some cases, driving the rain animal. One was copied by Stow and reproduced some eighty years later by Rosenthal (Rosenthal, 1953, plate 1). The rain animal has a hippopotamus-like appearance and the site is in the Harrismith district where, in the past, there would have been an abundance of hippopotami to

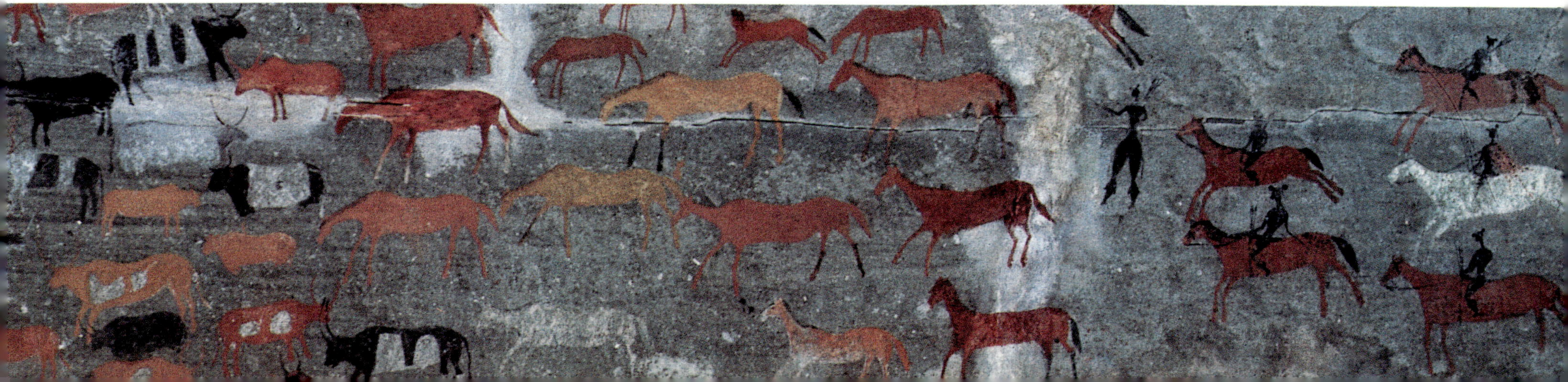

provide models. Its rain animal identity is established by the activities of the people near it and the cloud-like aura surrounding it (Figure 79). There is a similar scene within a few kilometres at a similar site and in a similar environment (Figure 80).

There is obviously a fine line that demarcates paintings depicting the leading/driving or capturing/charming of a rain animal and those reflecting an actual hunting situation, but it seems reasonable to assume that where the people in the scene are all armed and actively involved in pursuit, and the animal is a clearly identifiable species, particularly if wounded, the artist was painting an event that actually took place. In contrast, where people in the scene are engaged in dancing or show attitudes compatible with the trance experience and/or the animal is not clearly recognisable (especially if it has features signifying clouds or rain) then the scene is symbolic of the rain animal. Four examples that have the latter features are illustrated. The first is from Lady Grey, the second from Aliwal North, the third from Herschel and the fourth from the Transkei — essentially the same topographical region (Figures 84, 85, 86).

Of particular interest is an animal that was first published by Willcox (Willcox, 1956, plate 67). The painting is at Bamboo Hollow in the Giant's Castle Nature Reserve (Figure 90). First considered to be a cow under attack by hunters, it is now recognised as a rain animal summoned during trance experience (Lewis-Williams, 1983, 7). It is surrounded by figures in various positions associated with trance or with the 'hunting' of the conceptual rain animal but, most unusually it also has a line of crosses which are known from other paintings, in all likelihood, to represent bees. They are flying to or from the animal's shoulder. That is the spot where the power of the animal is concentrated. Bees are recognised by Bushmen as having great significance as they provide the most important source of sweetness and the basis for honey beer or wine. Therefore, their association with that particular part of the animal is not surprising as it is the point at which the power of the animal might be likely to leave the body in order to control the weather.

To the !Kung Bushmen, for example, that power is known as *n!ow* and is considered to be present in all people and all large animals, having been acquired at birth. Each person considers that his/her *n!ow* has a particular effect — good or bad — on the weather (Marshall, 1957, 234). A good effect is to bring rain, a bad one is to bring bitter cold.

A further example of the driving of a rain animal may well be provided by a painting copied by Stow and then lost for the best part of a hundred years until rediscovered by Neil Lee and I (Woodhouse & Lee, 1976, 29) (Figure 82). It has been considered to be a literal hunting scene of a hippopotamus pursued by hunters wearing hartebeest-head masks as disguises, but with both a rhinoceros horn and buck horns added to the hippopotamus at a later date. It is not a very satisfactory explanation in view of the animal's long legs and the white under its belly which is more prominent in Stow's copy and is likely to have worn away in the period prior to my photograph. There is little doubt that this is a rain animal being driven over the countryside, in spite of the opinion of Miss Bleek's informant that it represents a rhinoceros hunt (Stow & Bleek, 1930, plate 63).

Perhaps the most beautiful of the pictures illustrating the driving of the rain animal is on the Lesotho side of the Caledon River (End papers) where the attitudes of the men surrounding the animal are so elegantly portrayed. They leave little doubt as to their intention to move the animal forward rather than to kill it.

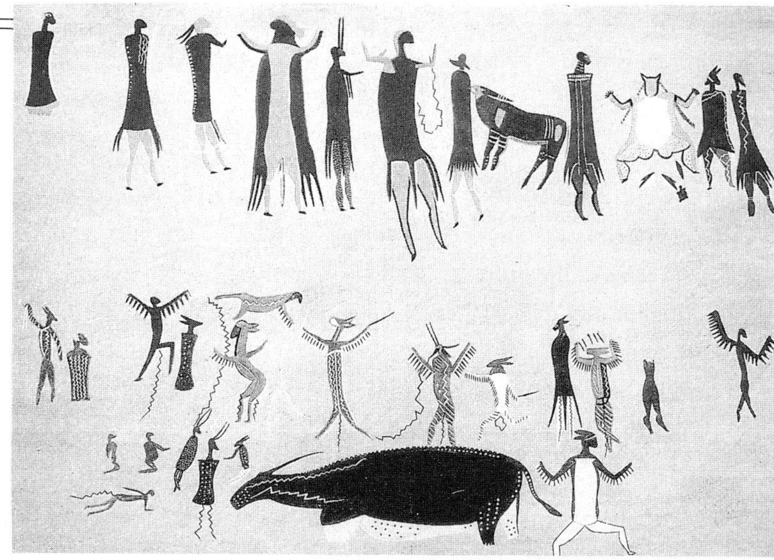

Figure 87

THE COMPLETE RAIN-MAKING SCENE

Overlooking Sani Pass, high in the Drakensberg, is one of the most complete and complicated of the rock paintings illustrating a rain-making ceremony (Figure 88). It was brought to the attention of Neil Lee and I by a young geologist called Van Heerden, and we described it under the name of Van Heerden's shelter (Woodhouse & Lee, 1978) but a subsequent publication (Maggs, 1979) let us know that it was already on record at the Natal Museum as 'Rain Dance Shelter'. So be it. A black-and-white photograph of the scene was reproduced with our article and also in *Bushman Art of Southern Africa* (Woodhouse, 1979, 103) but neither really conveyed the masterly technique and composition of this painting which can hold its own not only in these respects but also as an instrument of communication and record and as a reflection of a particular culture, with any work of art anywhere.

The composition consists of a number of units, each aesthetically satisfying and communicative in its own right, but linked to make a whole that is of special merit and interest. At the top a circular dance is portrayed, in the centre of which are medicine men or shamans flanked by women clapping. At the bottom are two rain animals, numerous fish, nets and men carrying coracle-style boats. On either side of the picture near the top are pairs of run-

Above: Stow's copy of paintings concerned with rainmaking removed from the Smithfield District to the National Museum, Bloemfontein by Charles Sirr Orpen.

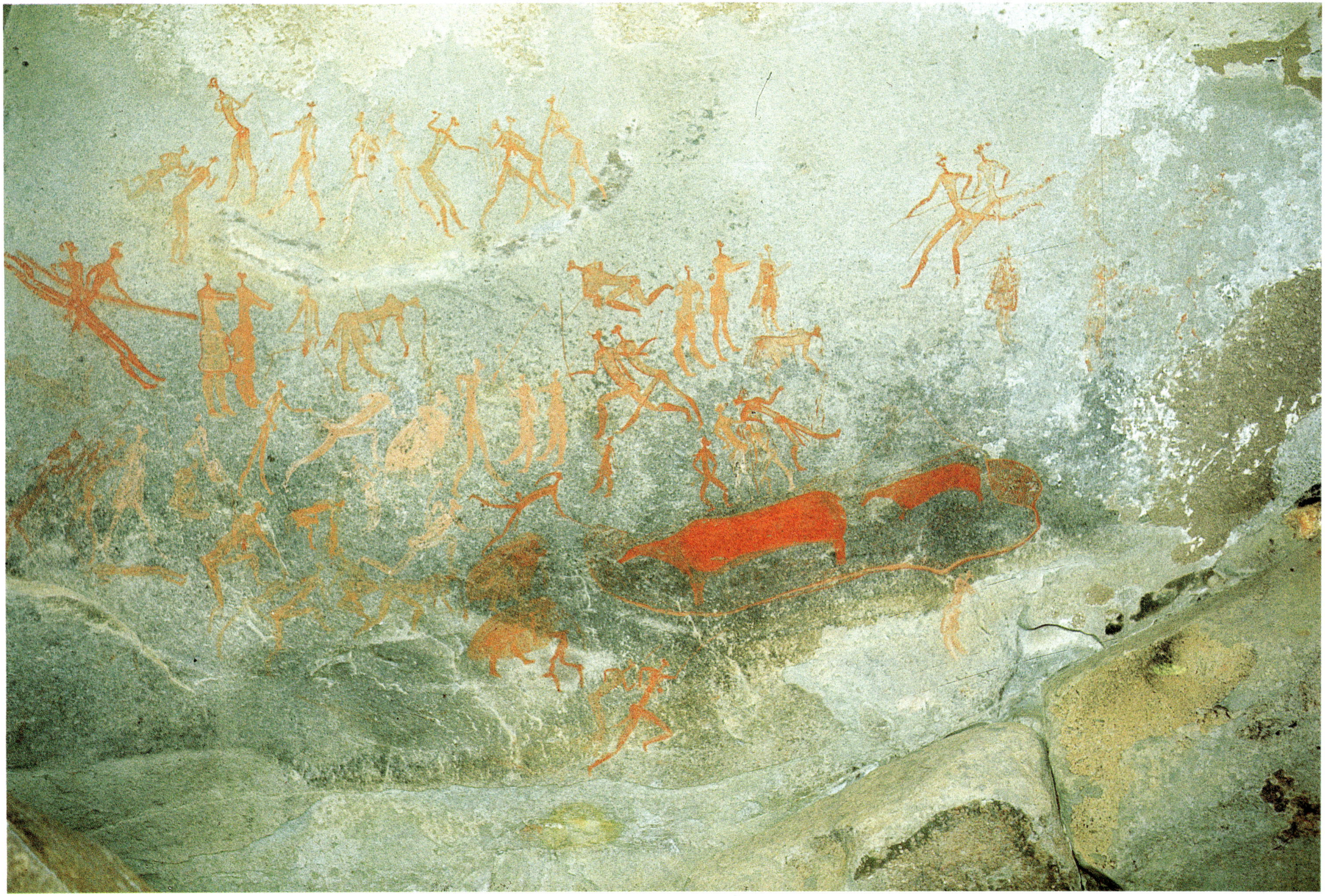

Figure 88: Complete rain-making scene with both Zulu and Bushman features. Underberg.

Assorted 'creatures of the rain' copied by Stow in the 1870s. There appears to have been little deterioration. Rouxville.

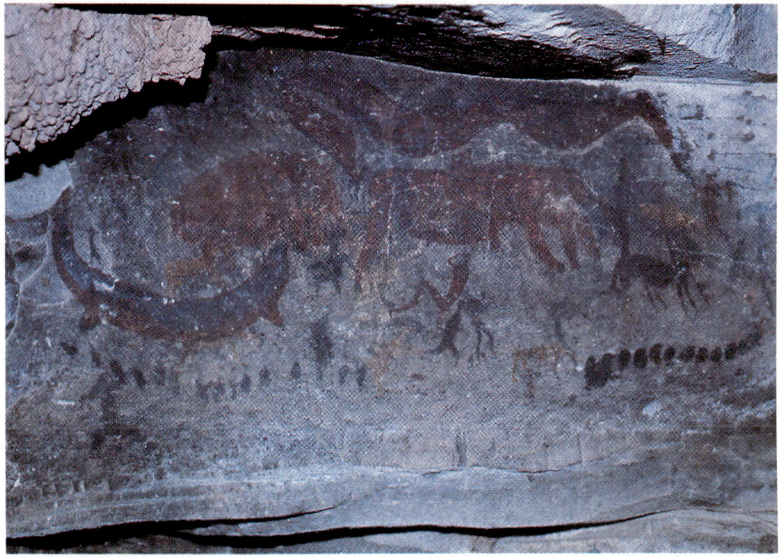

Figure 89

ning figures painted side by side — a device which Walter Battiss described as 'unique in prehistoric art' (Battiss, 1948, 224).

There are certain elements in this picture which appear to derive from Zulu sources although one can never be completely sure whether some customs, artefacts or items of dress used by the Zulu and Sotho peoples were not, in fact, obtained or copied from the Bushmen at an earlier date. However, attention should be drawn to the knobbed sticks carried by some of the men, the long-handled, small carrying bags of some of the men and women, the bladders in the hair of certain people, the head rings of two of the men and the skirt of animal tails worn by one of the rain doctors — all typical of Zulu material culture. The slender trunk of a rough-barked tree carried by one of the wide-striding pairs and the slender wands carried by the other may be explained by the custom of Zulu diviners who used special sticks for beating the ground and sent for rods for the same purpose (Krige, 1936, 300). Sending people out to search for particular objects needed in their 'medicine' or to look for things 'out of place' was an appropriate delaying tactic while making rain magic during a period of prolonged drought.

The nose-bleeding of the rain doctors is clearly depicted as is the prone position of the figure who has gone into trance. These features are essentially Bushman but the dance at the top of the painting is not typical of the trance dance and may be Zulu. The use of bladders in the hair certainly seems to be a cross-cultural feature as it is reported in both the Bushman (How, 1962, 51) and Bantu-speaking groups (Ritter, 1955, 90; Broster, 1981, 31). In this painting the wide-striding figures carrying the sticks or wands are wearing the bladders. They are likely to have been the acolytes of the doctors.

The observation above that coracle-type boats are carried by two people approaching the rain animals is supported by a painting illustrating the use of such boats for spearing fish which was reproduced by William Anderson in a report on the geology of Natal as far back as 1907 (Anderson, 1907). Other such paintings have been recorded (Vinnicombe, 1960). Such boats were quite likely to have been introduced to the Bushmen by one of the Bantu-speaking groups of the east coast which is renowned for its skill in the weaving of reeds, grasses and withes into nets, baskets, huts and even watertight containers (e.g. Knuffel, 1973). The well-known archaeologist, J Desmond Clark, studied the question (Clark, 1960) and also came to the conclusion that 'probably hybrid groups' were involved. The inclusion of the boats in this painting, together with nets and fish, illustrates a different food chain from those encountered in chapter 6. This one follows the course: rain — rivers — boats — nets — fish.

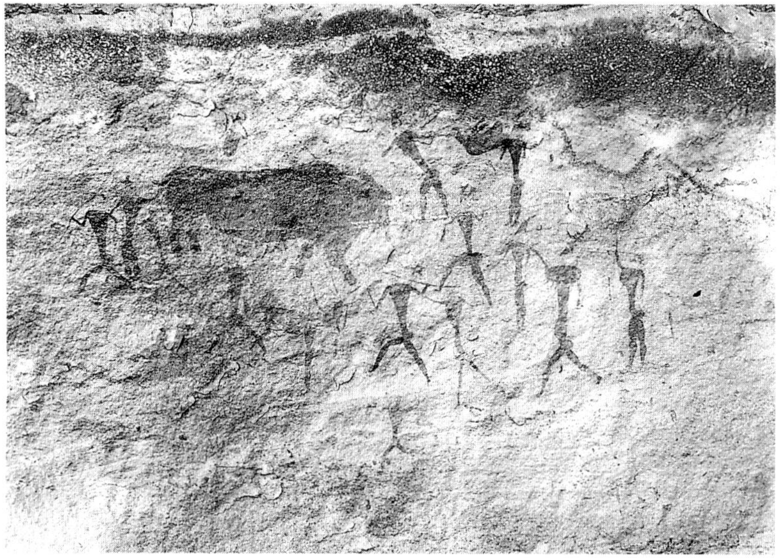

Figure 90: Leading/Driving the rain animal. Giant's Castle.

Compositionally not quite so satisfying, but originally extremely colourful, was a rain-making scene near the old French mission station of Beersheba in what is now the Smithfield district. I write 'was' because the copy made by Stow (Stow & Bleek, 1930, plate 58) (Figure 87) is beautifully coloured but the original, now in the National Museum, Bloemfontein, is a shadow of its former self. The museum authorities have done their best by providing a modern coloured copy for comparison and study but the masterpiece has almost disappeared.

The farm of Beersheba was sold by the French missionaries in 1858 to the father of a Mrs van der Merwe who told Dorothea Bleek that she remembered Charles Sirr Orpen frequently visiting her father and removing rock paintings — presumably those now in the National Museum. Stow's copy bears a note that the site was discovered by Chas. Sirr Orpen, brother of J M G Orpen, the magistrate who obtained much valuable information from the Bushman Qing.

When shown Stow's copy of the Beersheba painting by Dr Bleek and Lucy Lloyd their Bushman informant explained:

> Frog. Thought to be a girl who ate what she should not and was changed into a frog. Her people go to her. The rain is below, a black rain. It has killed the people. The girl ate *touken,* she displeased it. The arrows become reeds and stay at the spring, the sticks become bushes and are at the spring [house sticks I believe] (Stow & Bleek, 1930).

The 'frog', presumably, was the fat figure third from the right in the top row of figures in Stow's copy. If correct, this identification is useful as there are similar figures in other paintings. The identification of 'reeds', presumably painted in white between the fourth and fifth figures in the top row and upside-down, in the bottom row is also useful because it confirms the use of 'plant form' for a similar painting referred to in chapter 10. The explanation refers to a story known as 'The Girl's Story; The Frog's Story' (Bleek & Lloyd, 1911, 199-205) in which a girl ate the children of the Water (their identity is not known but they were striped and as big as calves) and was caught up in a whirlwind, followed by her parents and all their possessions, and they were changed into frogs and reeds respectively.

There is a wealth of symbolism in this painting; therianthropic figures, the partially spotted, black rain animal, zigzag body painting symbolising water and the fringed arms like wings which will be referred to in more detail further on. Unlike the painting dealt with at the beginning of the chapter this one would appear to be solely Bushman in concept with no elements derived from contact with Iron Age people.

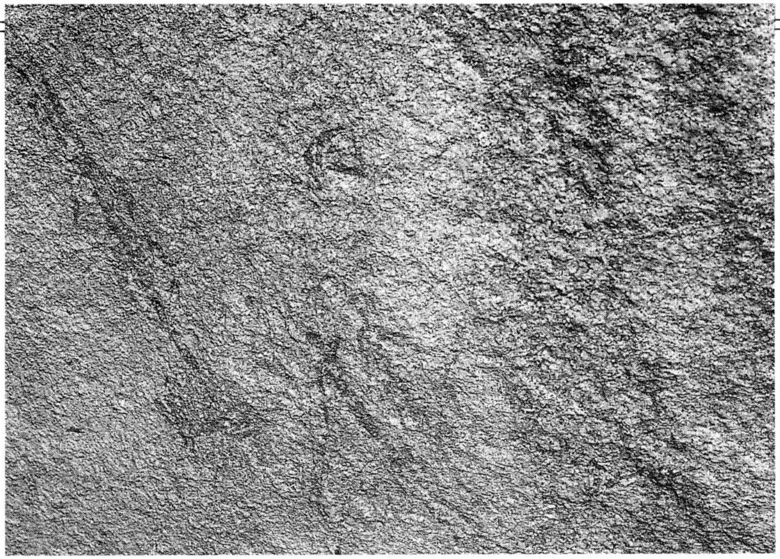

Figure 91

HUMAN SACRIFICE

The practice of sacrifice is not recorded anywhere in the literature dealing with Bushmen but the possibility that it existed was brought to light in chapter 3 which dealt with an offering to a rain snake, and again in chapter 9 when a buck impaled with a stake was observed in the sophisticated scene near Maclear. Both these paintings are from areas where there was contact with Bantu-speaking people over an extended period, probably of several hundred years.

The literature dealing with the Bantu-speakers contains many references to sacrifice. It is especially prominent in connection with rainmaking, extending from the sacrifice of a fowl to that of a bull and even to humans (e.g. Larson, 1966). The colour of all such sacrifices had to be black in order to produce black clouds (Dornan, *c* 1923).

In the same way that the rain-making scene from Natal reflected cross-cultural contact between Bushmen and Zulus, it is reasonable to assume that many of the Zimbabwean paintings would have reflected a similar contact between Bushmen and groups ancestral to the Shona. This would seem to apply to a unique painting in the Marondera (formerly Marandellas) area of which Mrs Goodall wrote:

> A female body is lying on the ground, the person is sacrificed in time of drought, to induce the longed-for rain. A tree grows out of the body . . . When the tree reaches the clouds the rain goddess, towering on high, pours out the desired rain . . . Figures creep up the tree which, in the upper parts, is transformed into a snake with ears . . . This mythic tree has been called the Tree of Life . . . (Goodall, 1959, 102)

Above: Detail of a painting of a sacrifice as part of a rain- making ceremony. Marondera (formerly Marandellas), Zimbabwe.

A sacrifice as part of a rain-making ceremony. Marondera, (formerly Marandellas), Zimbabwe. The drawing is a reproduction of the painting on the left.

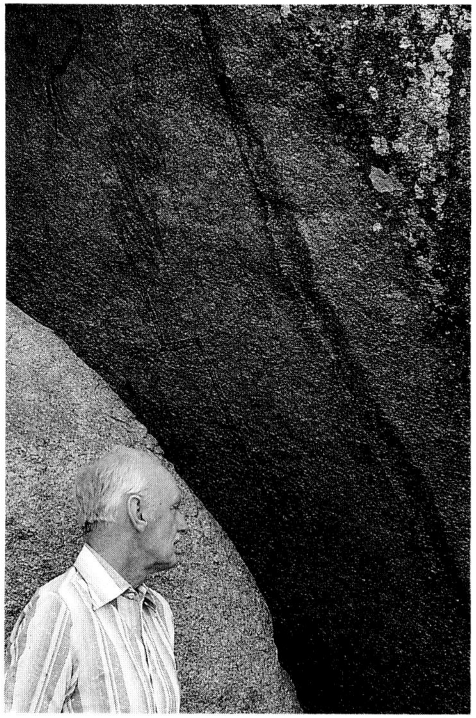

Figure 92

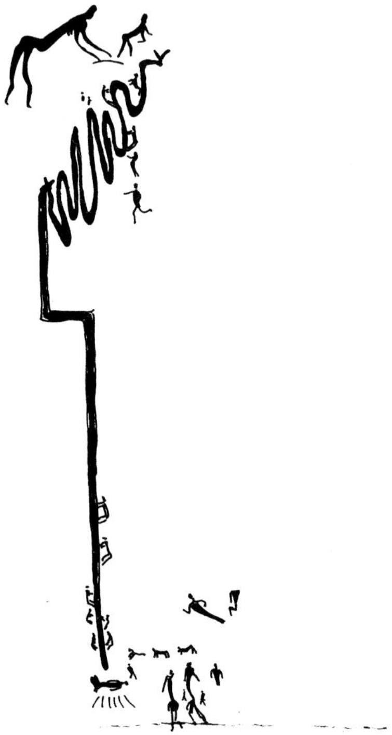

Figure 93

The painting is almost impossible to photograph without distortion owing to its situation in a narrow cleft between two large boulders but there is no doubt that it was an extremely important painting and may well have depicted human sacrifice (Figure 92). Yet the word 'tree' is not really substantiated. There are many paintings of trees in Zimbabwe and they all look much more like trees than in this painting. I have been unable to find any recorded information on rain goddesses in either Bushman or Shona mythology. It seems more likely that the female figure at the top of the painting is an additional representation of the victim at the bottom, indicating that the ceremony has been successful and that a rain snake has been summoned up to create a river of water. But there are also difficulties with this interpretation. What is the significance of the right-angled turn in the 'river' followed by a series of very sharp zigzags just before the head is reached? Why are figures on all-fours ascending the 'river' and zigzags? Are they there to indicate participation in a trance dance, thus emphasizing the conceptual nature of the whole scene? Are the zigzags an example of the patterns seen in the mind's eye during trance?

Regrettably, I cannot answer these questions but can only compound the problem by indicating that there are other examples of 'buck-headed snakes/rivers' and associated 'all-fours people' at several sites in Zimbabwe including the well-known site of Glen Norah near the Harare sewerage works where they appear to be kicking backwards with their legs while their hands rest on what Mrs Goodall has described as 'rope bridges'. Bridges and rivers go together but this may well be an unduly ethnocentric explanation. It is also difficult to accept that people would cavort in what Mrs Goodall describes as 'comical attitudes' while crossing a bridge. Elsewhere Dr Lewis-Williams has suggested that rather similar lines to those making up the 'bridges' represent the circular groove worn by the feet of trance-dancers. There is obviously much more field-work and study to be done before it can be said that we really understand these interesting paintings.

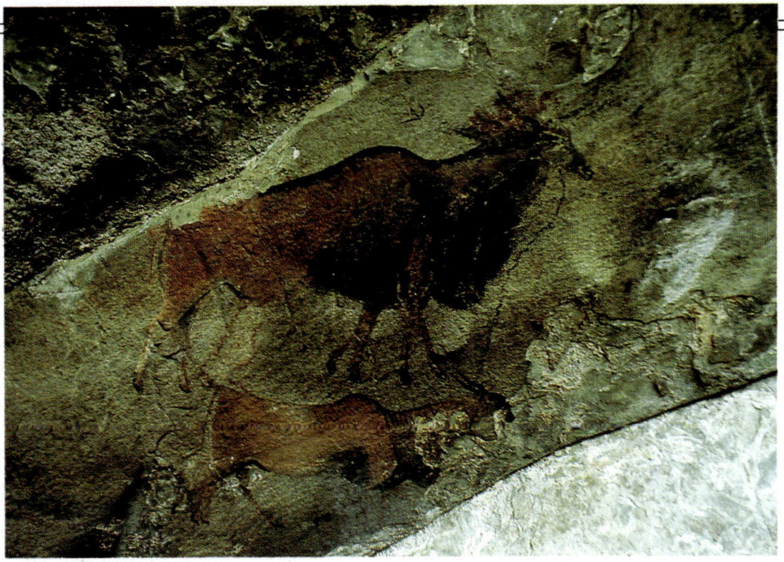

Figure 94

THE ELAND AS RAIN ANIMAL

In the Drakensberg and nearby regions there are so many paintings of eland that the casual observer might get the impression that the artists painted little else. Research has confirmed the numerical superiority of the eland over other animals, although it exceeds the rhebok and unidentifiable buck by a comparatively small margin (Pager, 1971, 327). The eland also features very prominently in the huge corpus of ethnographic material collected by the Bleeks, Lucy Lloyd and Orpen. As a result, two significant works interpreting the meaning of the paintings, *People of the Eland* (Vinnicombe, 1976) and *Believing and Seeing* (Lewis-Williams, 1981) have both devoted considerable attention to the relationship between the eland and rain.

It is not proposed to cover in detail the ground that has already been traversed so comprehensively, but it is surprising that recognisable relationships between paintings of eland and water or rain are not numerous. The accent in this statement may well belong to the word 'recognisable'. It is quite likely that many paintings of what appear to be uncomplicated representations of eland, singly or in herds, were intended to communicate feelings and beliefs concerned with rain. For the present, however, it is proposed only to note those paintings where literal representations of eland are obviously complicated by more abstruse concepts. One of the most interesting has only recently been recorded in the Bethlehem District (Dowson & Holliday, 1989). A group of red monochrome eland, one of which is upside-down, is painted over a vertical screen of zigzag lines. Other lines representing liquid stream from the nose and rump of the upside-down animal (Figure 95). While the authors of the report prefer, for well-argued reasons, to regard these eland as depicting the release of supernatural potency, to me they are one more example of the large corpus of art depicting rain animals, of which

Above: Rain eland. Clocolan.

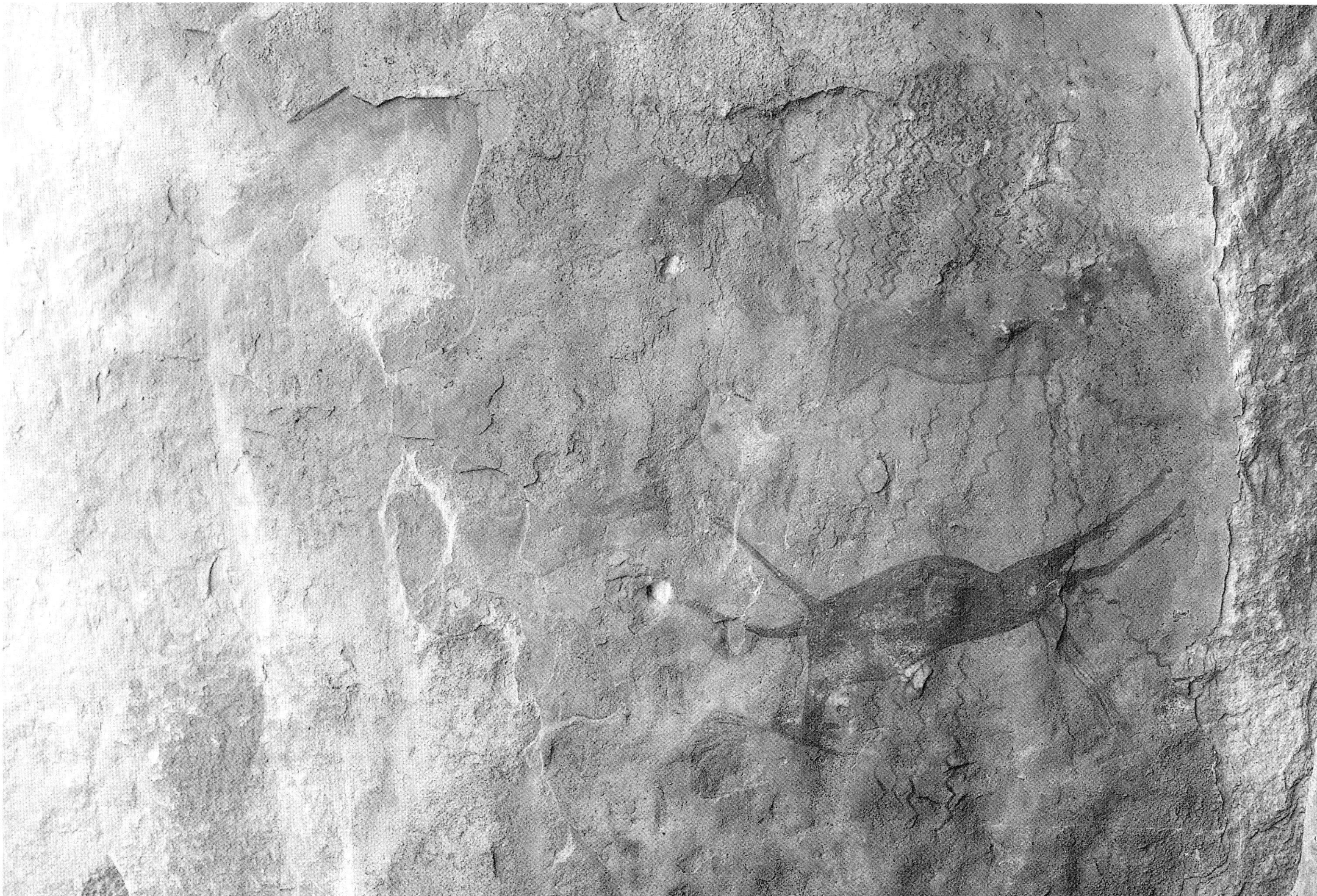

Figure 95: Group of rain eland from the same district as the prototype rain animal. Bethlehem.

Figure 96: Therianthropes (partly human, partly animal) which form part of a rain-making scene. Tsoelike Valley, Lesotho.

Figure 97: Rain eland with rain snake. Harrismith.

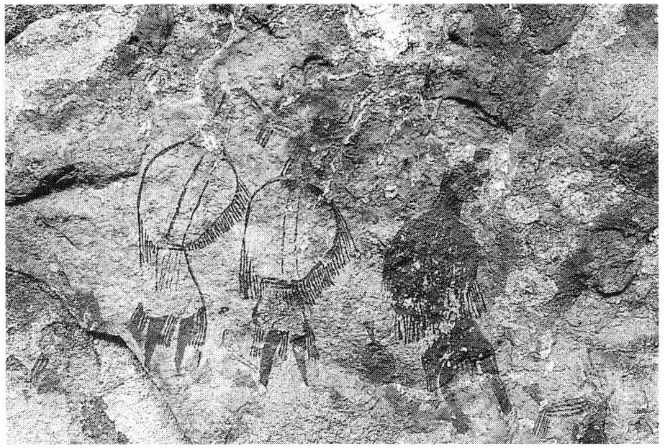

Figure 96

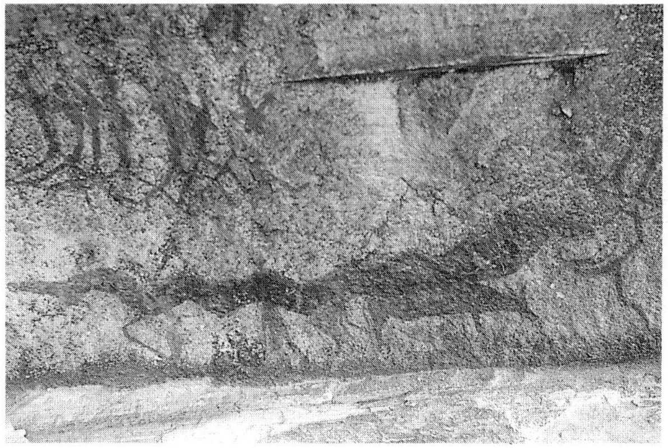

Figure 97

there are so many other examples in the Bethlehem and adjacent districts. I photographed them in company with Neil Lee in 1986.

Another excellent example (Figure 98) was reproduced by Vinnnicombe (Vinnicombe, 1976, Figure 237) with a descriptive caption and a reference in the main text to 'game sorcerers'. I used part of the same painting to illustrate an article in which I first drew attention to the linear feature which links many animals and people in the paintings. It consists of a red line edged with white or cream dots and is obviously a significant concept which may be regarded as a 'line of magic force' (Woodhouse, 1975) or as 'supernatural potency' (Lewis-Williams, 1981b) which amounts to much the same thing. In this scene the line emerges from the nose of the eland and passes to a human figure wearing a white garment, and then round the chest of a second figure with a headdress consisting of the bladders that distinguish a medicine man. The eland could be seen as being 'led' by the magic force of the two human figures and the whole picture — which includes a procession of twenty-three (Figure 96) kaross-clad, animal-headed figures to the left of the eland — could depict a rain-making ceremony. In her caption Vinnicombe describes the creature above the eland as 'an anteater-like animal with stripes on its face' but it seems more likely to be an eland depicted as viewed from behind, like many similar paintings of eland with lines on their faces in other parts of the Drakensberg. A very clear eland accompanied by another example of the 'wavy line' motif appears in the Molteno District at the same site as the picture shown in Figure 62.

A large and impressive animal with the general appearance of an eland but with its front legs firmly wedged between shapes that resemble the huge piled rocks of that part of Zimbabwe, may be seen in the Lake McIlwaine Nature Reserve near Harare (Figure 100). Mrs Goodall captioned her reproduction: 'Animal trapped between rocks' (Summers, 1959, plate 8) but the presence of trees, bees and people in the same scene makes it more probable that it is a representation of a rain animal above a rocky landscape.

In his book *Believing is Seeing* in which he draws attention to the close relationship between eland and rain, Lewis-Williams reproduces a drawing of what he considers to be a rain eland from the shelter in the camp at the Giant's Castle Nature Reserve in Natal. For the record this shelter has, in the past, been referred to as Snake Shelter. His reason for associating this particular eland with rain is that it is surrounded by the 'winged figures' which he refers to as 'trance buck' and which have been called 'flying buck' and 'alites' by other authors (Lee & Woodhouse, 1964; Pager, 1971). Its association with the trance buck may not necessarily imply that it is a rain eland. In this instance the medicine men in trance represented by the trance buck may

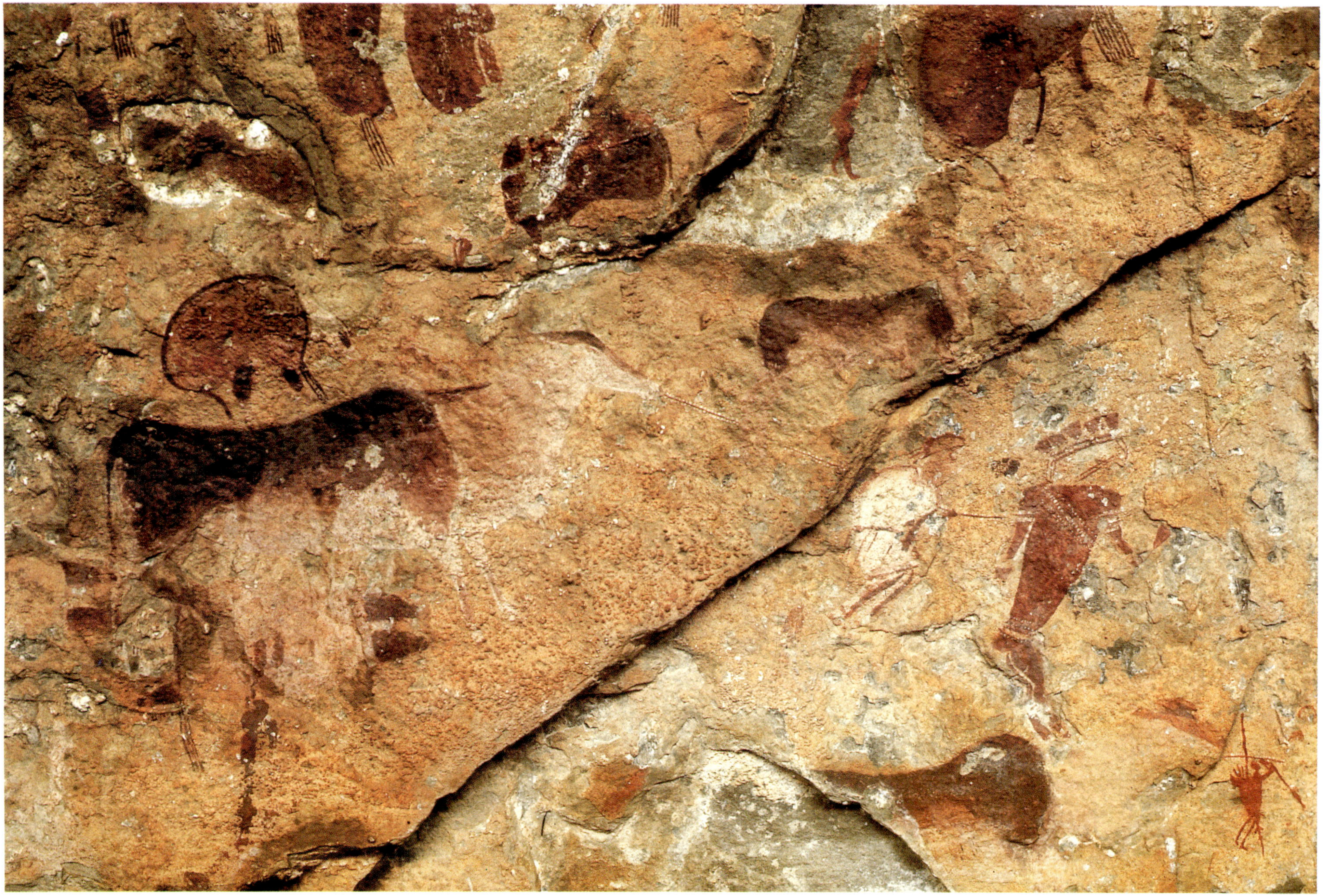

Figure 98: A 'line of magic force' connects an eland (a rain eland?) with a medicine man or shaman. Above the rain eland is another painted as seen from behind. Tsoelike Valley, Lesotho.

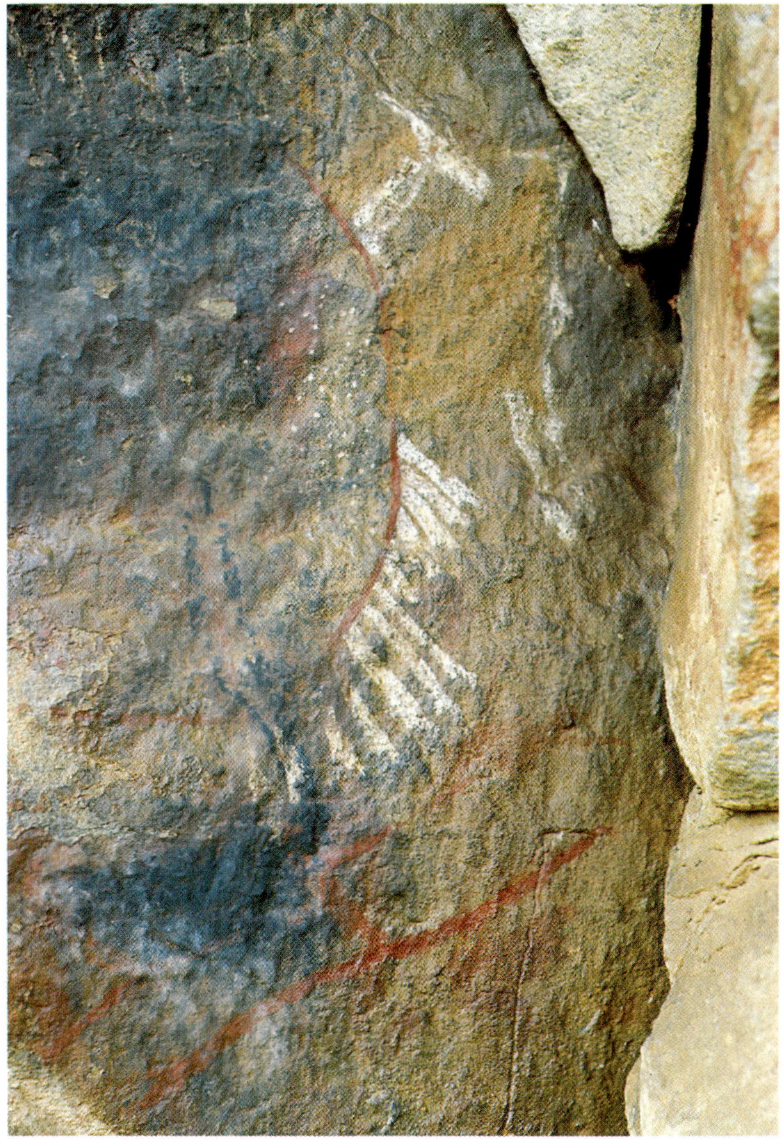

Figure 99: Herd of rain eland emerging from rock face. Harrismith.

have been involved in ceremonies to increase the number of eland rather than to bring rain. There were game sorcerers as well as rain sorcerers.

A rather more likely candidate is the eland in the painting in the Harrismith district which I first used to illustrate the suggested practice of 'playing with eland' based on the well-known taunting behaviour of Bushman hunters during the last hours or minutes of a dying animal (Woodhouse, 1971 as modified in 1979) (Figure 97). This particular animal has a line hanging from its neck and an unnaturally long tail which is grasped by a human figure while another stands on one tiptoe with arms thrown back facing it. The tiptoe figure is in an attitude associated with the trance dance and the line from the neck of the eland could be seen as the means of leading the rain animal.

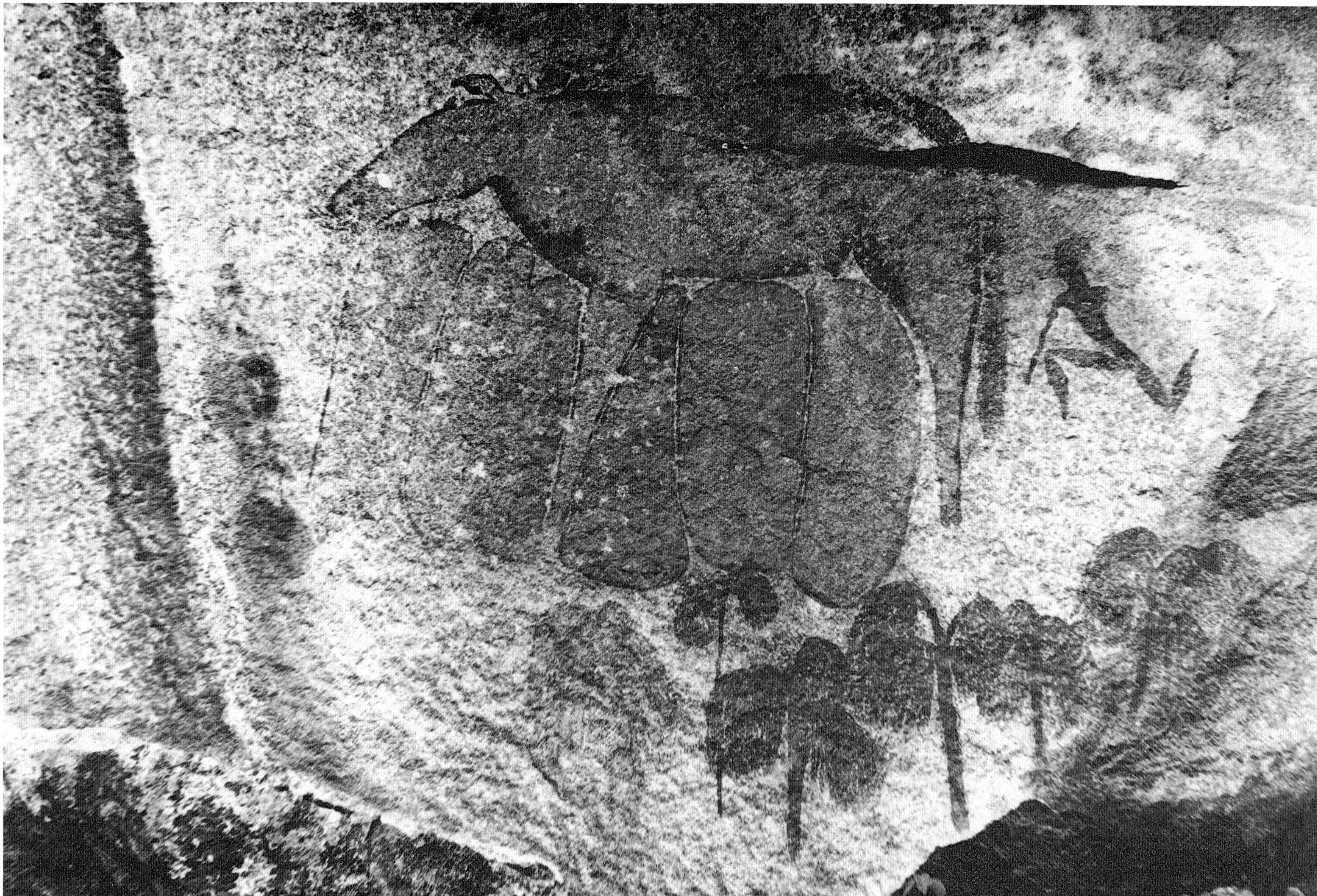

Figure 100: Rain animal, modelled on an eland, on the rocks. Typical of the guti *or clammy mist that swirls across the Zimbabwean landscape from time to time. Trees are more frequent in paintings in the Mashonaland area than elsewhere. Lake McIlwaine.*

The claim that this painting depicts a rain animal is strengthened by the fact that there is a snake partially superimposed upon it and that there is an accompanying group of striding human figures somewhat similar to that discussed in chapter 3. There are no showers falling from either the eland or the snake in this case, but at the site near Newcastle recorded by the missionary Prozesky in the 1870s and rediscovered by Robbie Steel and I with the assistance of Harald Pager a hundred years later, there is an eland which is also the centre of a tail-pulling and confrontational situation. A man hangs from the eland's neck and from the dewlap there falls a cluster of lines that could certainly depict a shower of rain. Again, I have previously described this scene as a literal representation (Woodhouse, 1972, 92) of 'playing with eland' but the alternative explanation of 'rain eland' seems preferable as there are also some rather 'lumpy' animals of indeterminate species in the scene which may be regarded as rain animals.

The question of whether there may be more than one rain eland in a painting is raised when viewing an interesting painting of a group of eland in the Clocolan district. They are connected by red lines at the back and a wavy black line at the front (Figure 94). These lines are similar to lines in the Stow copy (Figure 87) discussed previously. The upper eland appears to have two sets of legs so it probably represents two animals walking side by side. In other words, there is a total of three animals. This painting contains no 'showers' but the wavy lines are taken as being representative of rain animals. Above the upper eland is a tiny black wide-striding hunter which, if it gives a scale, indicates that the eland were visualised by the artist as very large, strengthening the interpretation of rain eland. In several other cases, e.g. at Bellevue and Cat Cave (Willcox, 1963, plate 33; Woodhouse, 1978, 12) there is a similar disparity in sizes but no other indication that these are rain eland rather than literal representations of scenes of eland-hunting.

Near Harrismith in the Orange Free State there is a painting of an eland's head and neck, the front half of another eland and a cluster of eland legs, all of which emerge from behind a red line which edges a dark mass spotted with white (Figure 99). The eland are essentially herd animals. This painting could well represent an eland herd in its role as thundercloud sending a fierce 'he rain' represented by the legs of the eland and the white spots of hail. The 'he rain' stamps its spoor deeply in the ground (Marshall, 1962, 234) so such a group of legs and hooves would have been an appropriate representation of a heavy rainfall. The apparent 'emergence' of the animals from a cloud depicted in this painting leads to another line of investigation.

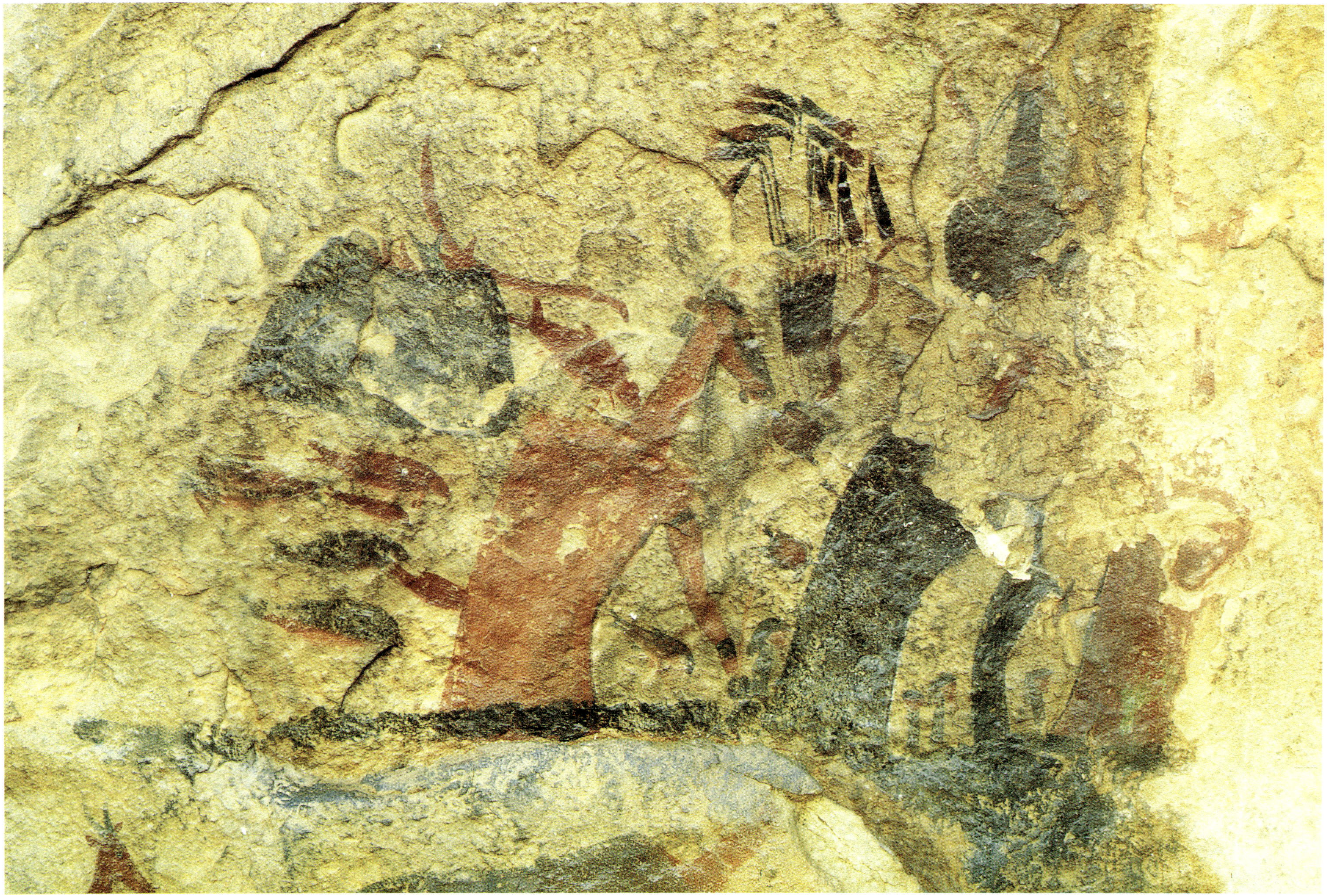

Figure 101: 'Creatures of the Rain' — eels, fish and tortoises with a hartebeest-headed therianthrope emerging from a line painted on the rock. Lady Grey.

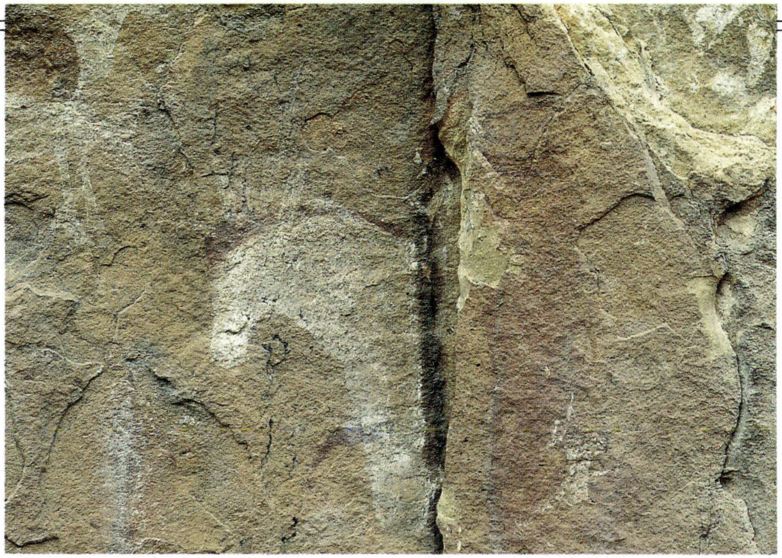

Figure 102

RAIN ANIMALS 'EMERGENT'

The eland and the cluster of legs noted at the end of the last chapter were given the appearance of emerging from a mass of cloud by being cut off by the red line and the black paint to the left of it. A rather similar technique was used at a site in the Lady Grey district where a black line cuts off the lower parts of three therianthropes of which one clearly has the head of a hartebeest (Figure 101). This painting was first recorded by Walter Battiss (Battiss, 1948, 160). The association with rain is established by the eels, fish and tortoises, all 'things of the rain' (Bleek, 1933, 303) which are painted close to the hartebeest-headed creature. It is associated with several other therianthropes and a large herd of eland.

The cut-off effect obtained by using the painted lines at Harrismith and Lady Grey was achieved in a different way to the technique of painting heads emerging from a crack or overlap in the rock face at Maclear (Figure 50) and Wodehouse (Figure 53). The same technique was used near Slabberts in the Orange Free State where the head and neck of an eland (Figure 102) emerge from a crack in the rock and at another site not far away, where a rain snake with a fearsome head convolutes in and out of a similar crack.

The use of a natural feature to enhance the painting's effect has only recently been recognised as a widespread technique in the rock art of southern Africa (Lewis-Williams & Dowson, 1990; Woodhouse 1990) but it is clear from these examples that it was intentional. The technique accentuates the sophisticated thinking of the artists who employed it, and it is also probably associated with the reported ability of the rain medicine men to travel below ground while in trance (Biesele, 1979).

Above: Eland 'emergent'. Fouriesburg.

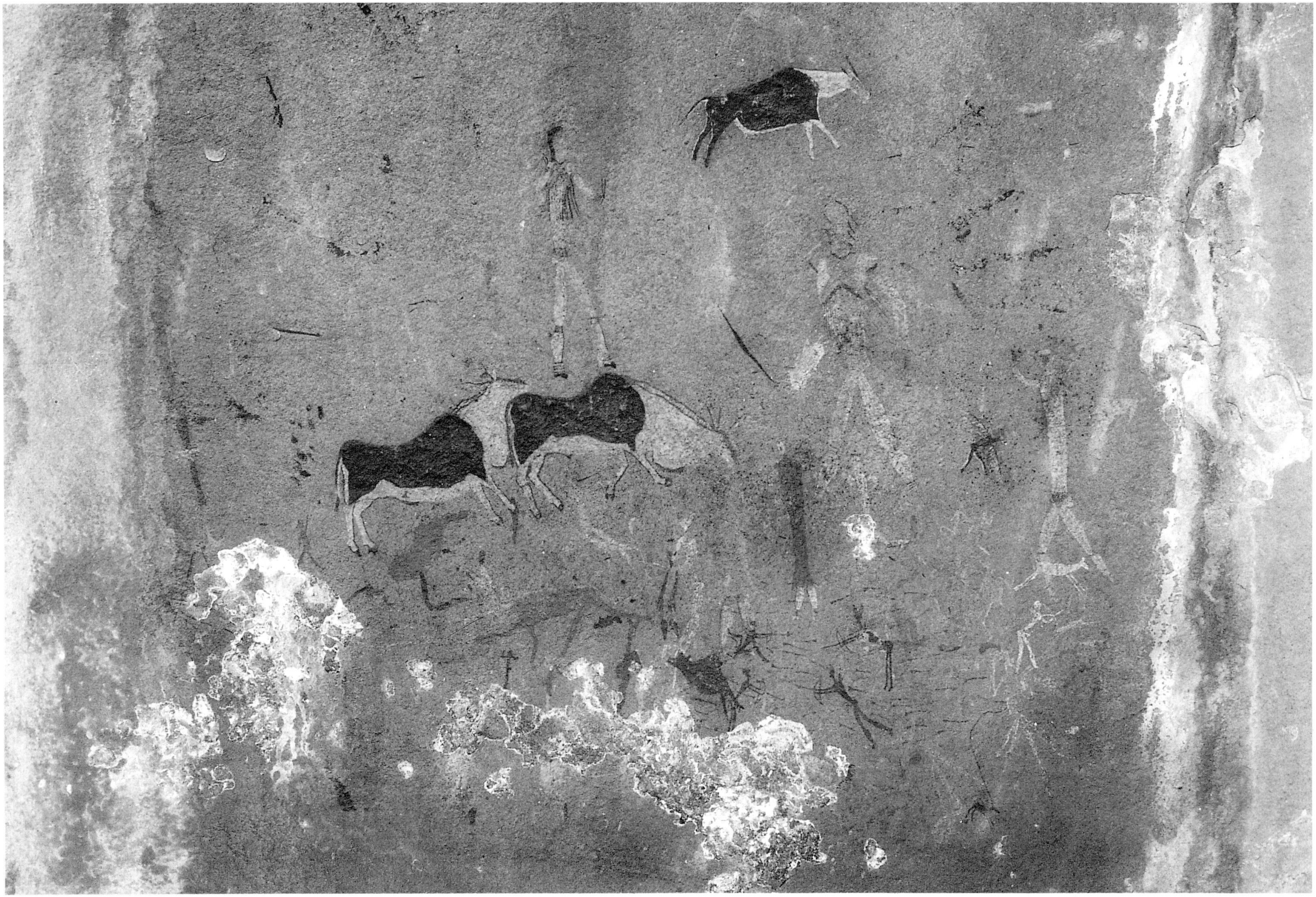

Figure 103: Rain sorcerers or shamans, one with beaked mouth. Aliwal North.

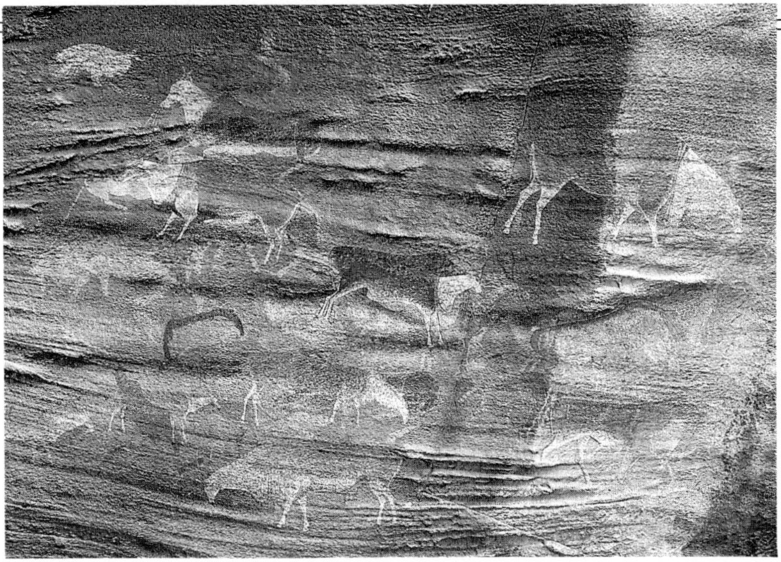

Figure 104

RAIN BIRDS

Two large birds cover nearly a metre of rock face at Markwe in Zimbabwe. Their nearest living counterpart is the ground hornbill but it has been suggested (Thornycroft, 1978) that they are rain birds. Mrs Goodall's copy (Summers, 1959, plate 10) supports that contention as it shows white streaks from the breast of one of the birds (these are not, however, visible on my photograph) (Figure 107). In the Barkly East district a white and red bird is joined to a herd of eland by lines of magic force and it has been suggested (Battiss, 1948, 212) that it represents the 'lightning bird' of Xhosa mythology (Godfrey, 1941, 2) (Figure 104).

There are many paintings which include winged figures, notably that from the old mission station at Beersheba in the Smithfield district (Figure 87) which was dealt with as a complete rain-making scene in chapter 14. When encountering winged figures in this context, it is invariably noticed how the feathers hang vertically downward from the wings in the manner of streaks of rain. If this suggested symbolism is correct, then it helps to explain a considerable number of previously unexplained paintings. One of the best known is that at Nsangwini in Swaziland which has appeared on the postage stamps of that country. The winged figures in this example have long 'beaks' which accentuate their bird-like appearance and they are accompanied by a procession of men carrying aloes (Woodhouse, 1979, 68) which normally bloom in the dry season: the appropriate time for a ceremony to ensure future rain.

There is a similar painting of figures with their wings extended sideways

Above: Walter Battiss suggested that the white bird in the top left corner was the lightning bird of Xhosa mythology. It is connected with the herd of eland by lines of magic force — or supernatural potency. Barkly East.

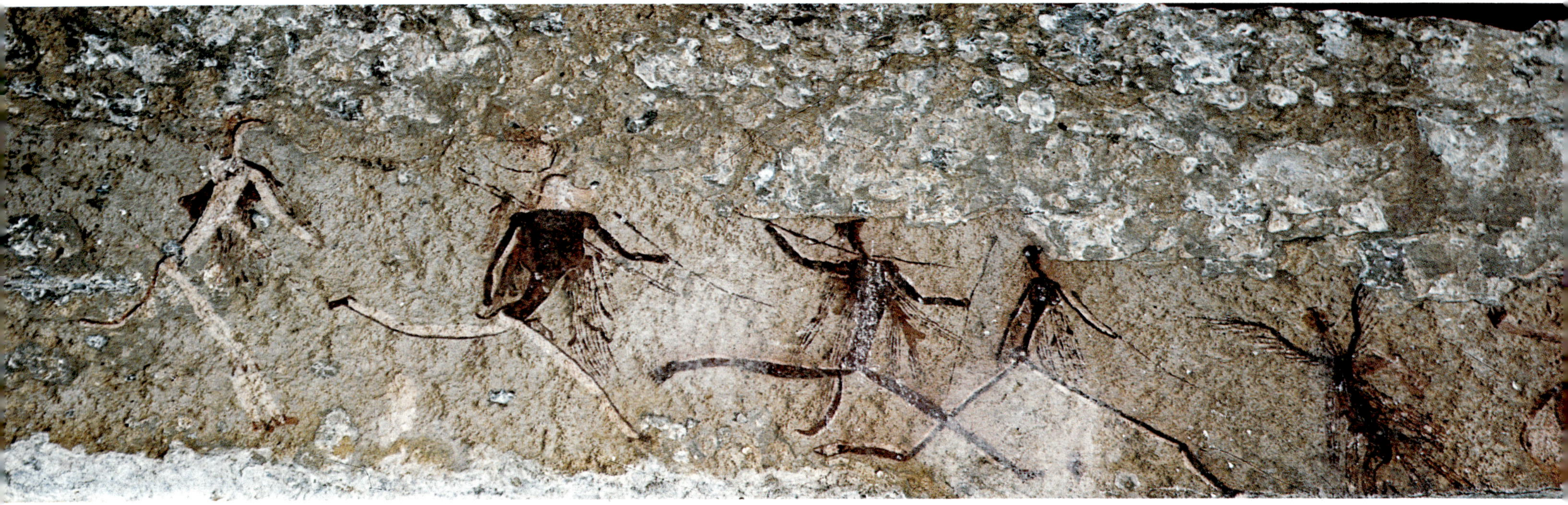

Figure 105: Therianthropic figures associated with rain. Barkly East, now in the Africana Museum, Johannesburg by whose courtesy it was photographed in 1969 and is now reproduced.

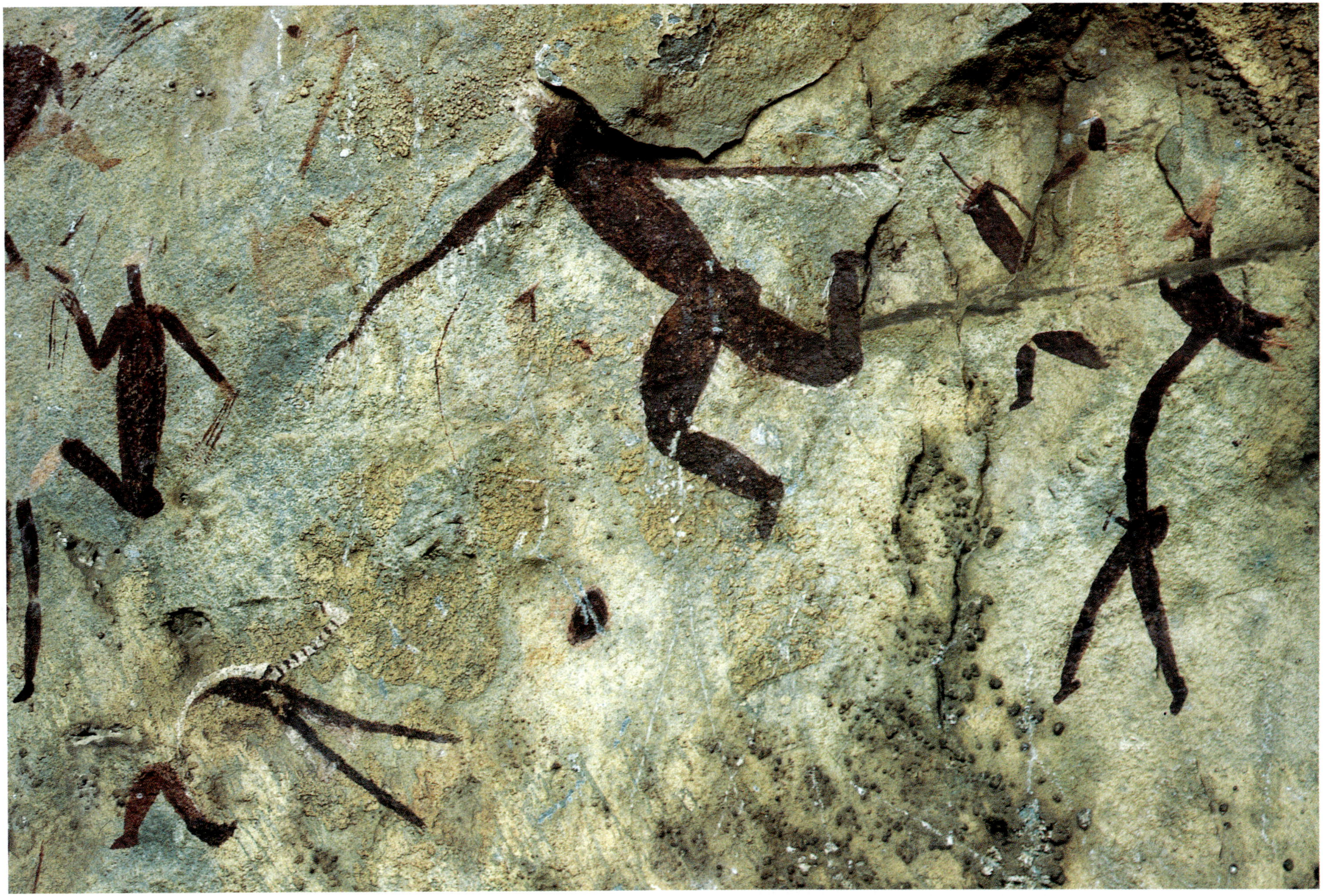

Figure 106: Winged creatures. Mount Currie.

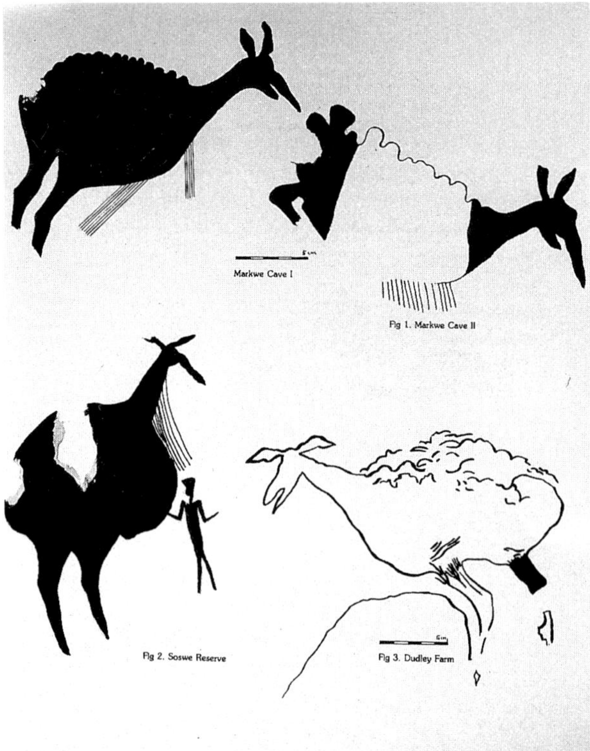

Figure 107: Fragmentary remains of rain birds. Reproduced by kind permission of Corona Thornycroft. Zimbabwe.

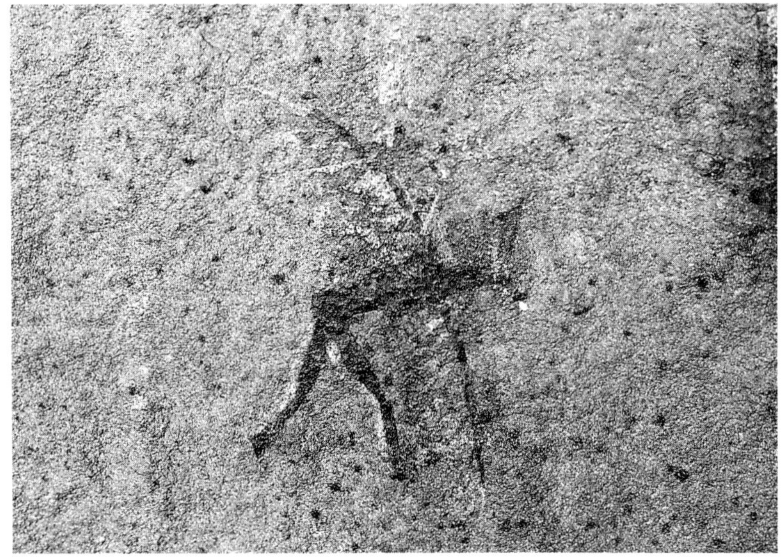

Figure 108: Winged Dancer, Aliwal North.

at Oatlands outside Harare (Goodall, 1959, 88). The figures have heads and horns like wildebeest and are associated with cheetah (Woodhouse, 1984, 25), thus linking them to the two fastest runners on the veld: appropriate when rain was needed in a hurry.

An elaborately painted scene from the farm War Trail in the Barkly East district was removed by Walter Battiss to the Africana Museum, Johannesburg, for safekeeping. It essentially portrays a group of varied, running therianthropic figures of which one has wings and several of the others have a plentiful supply of fringing, or showers of rain, falling from their armpits (Figure 105). The figure on the extreme right has the head of a bushpig or warthog and it may be of interest to recall that the wild pig is the totem animal of the family of Mojadji, the Rain Queen of the Lovedu people of the Northern Transvaal (Krige, 1980, 89). Both the birds and the bushpig, which are associated with rainmaking, seem likely to be further examples of cross-cultural contact. The rain sorcerer with a warthog head also comes to mind.

There is a winged figure in a scene at Kalkoenkrans in the Aliwal North district (Figure 108), first reported by Lynn Baard then of the National Museum, Bloemfontein (Baard, 1970), which has much in common with those at Nsangwini, Oatlands and War Trail. It is associated with white-painted figures (Figure 103) which have 'showers' falling from their armpits. One has the long, protruding, beak-like mouth which is also depicted in the Stow copy of the Beersheba painting. It would be appropriate for rain doctors to be depicted with beaks if they were engaged in rites involving rain birds.

Patricia Vinnicombe recorded and described a scene which included winged figures in a variety of attitudes in a high, windy shelter in the Drakensberg (Vinnicombe, 1976, 241, 2). The scene is adjacent to another which depicts a herd of eland and several alites or flying creatures and it also includes men in the 'arms back' position associated with the trance dance, as well as two rain animals. Two of the winged creatures are illustrated here (Figure 106).

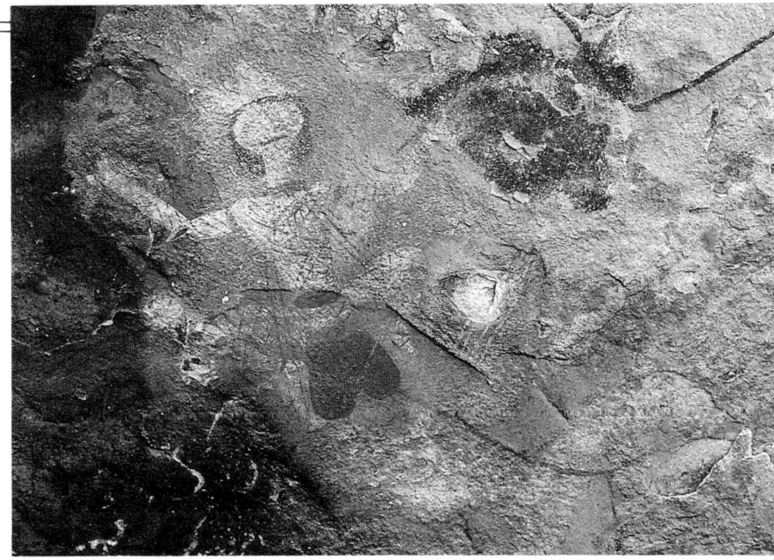

Figure 109

HUMAN 'SOURCE FIGURES'

More than a quarter of a century ago at the Fourth Pan African Congress on Prehistory, Mrs Goodall of what was then Southern Rhodesia, and is now Zimbabwe, delivered a paper on a distinctive mythical figure appearing in the rock paintings of that country, particularly in the Mtoko district. She had previously written an article on the same subject (Goodall, 1949) and her views were mentioned briefly in chapter 12.

The figure in question is essentially a fat figure, usually female (although both Goodall and I have recorded male examples) (Figure 110), and frequently with two long, wavy 'streams' from between the thighs (Figures 111, 112). Along these 'streams' human figures cavort in much the same way as has been observed of similar figures at other sites in Zimbabwe. Mrs Goodall suggested that the fat figures might be a local version of the universal 'mother goddess', but in subsequent discussion two important observations came from the floor. The first was from Professor Desmond Clark who was reported as saying:

> They suggest to me rather a possible connection of a divergent nature with the bushmen [sic] rain magic. That is a very well-known thing which is represented in the rock painting, the main characteristic being mythical monsters, sitting in a pond or connected with a pond with zigzag lines usually, representing water around them. I suggest it may be connected with that and also the representation of pythons with human figures on them again connected with rain magic.

There, expressed in the slightly quaint English of the French editor of the proceedings of the congress, is a link between Mrs Goodall's 'distinctive mythical figures' and the rain animals and rain snakes which are dealt with

Above: The source figure from Sorcerer's Rock, Ndedema Gorge, Natal, Drakensberg made famous by Harald Pager.

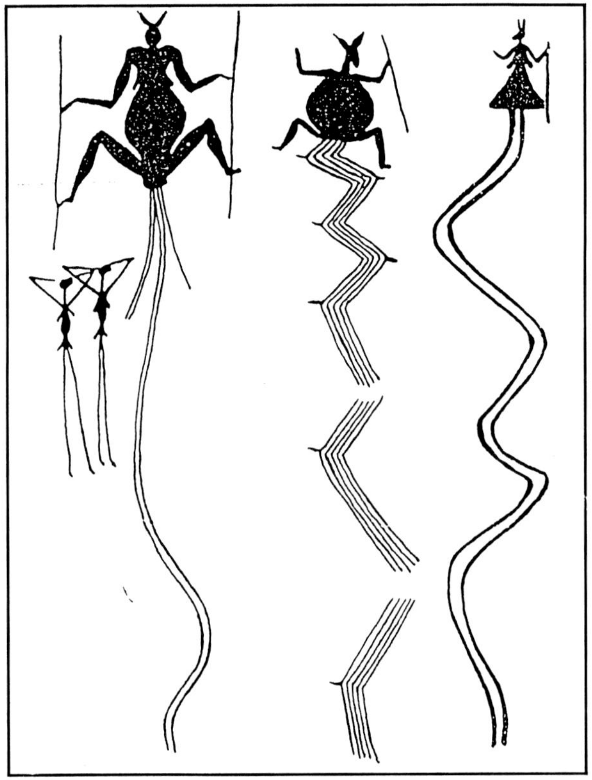

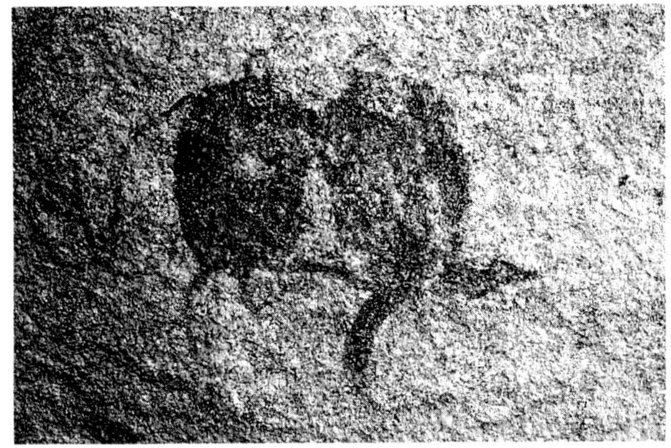

Figure 110: Male examples of the 'fat figures' found in Zimbabwe. Lake McIlwaine.

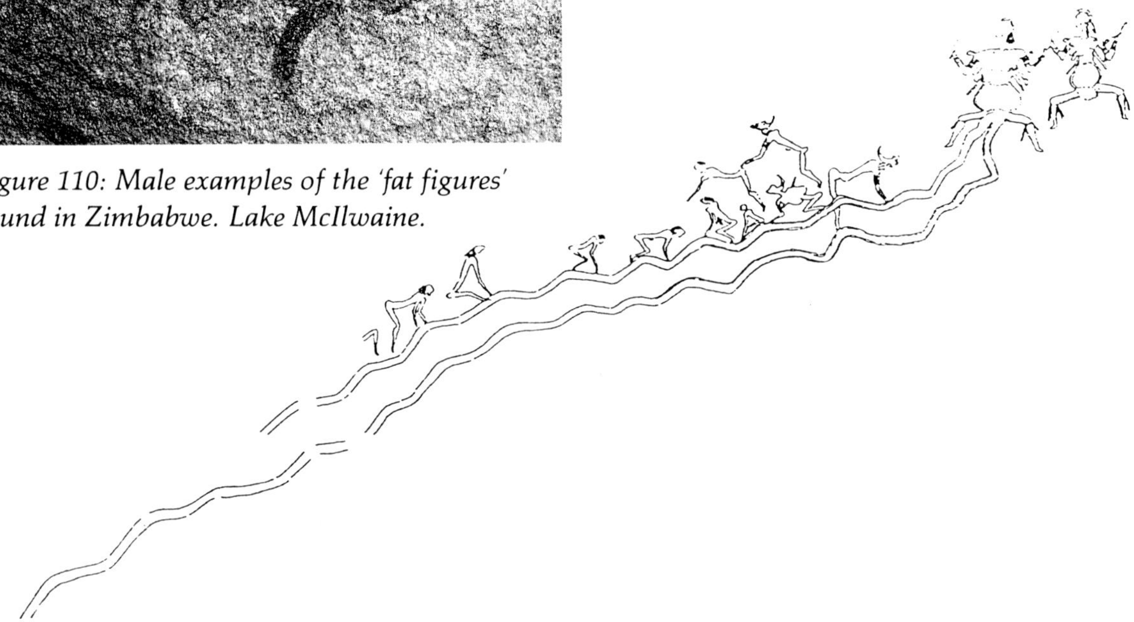

Figures 111 (above) and 112 (right): The figures recorded and written about by Elizabeth Goodall. Regarded as 'source figures'. Mtoko, Zimbabwe.

as the focal point in this book.

The second contribution from the floor was from Professor Phillip Tobias who was reported as saying:

> . . . might not [it] be some grossly caricatured representation of the macronympha, or enlarged labia majora of bush and hottentot and other peoples: two long skin folds that hang down between the labia majora.

The two contributions are not mutually exclusive. In fact, they are complementary if one regards the figures as 'source figures': similar to rain animals but in female form. Harald Pager drew attention to a similar figure at Sorcerer's Rock in the Ndedema Gorge of the Drakensberg (Figure 109) and a not dissimilar figure has already been discussed in the Stow copy in the Smithfield district — designated by a Bushman as a frog (Figure 87). There are two further examples in the Rouxville District at the same site as the 'master of the water' reproduced in chapter 11. One of these is painted in white, the other in dark red with a white outline. Both are spotted with white spots and in their vicinity are other spots, a white animal of indeterminate species and a dark hyena-like animal with a cluster of fine lines from its chest — all likely symbols of rain.

The lines associated with the 'source figure' at Sorcerer's Rock bear such a strong resemblance to a map of the local river system of the upper Ndedema that I have written about them in this connection elsewhere (Woodhouse, in Ms). To see them as rivers of water and also as a map of a river system is to see two sides of the same coin.

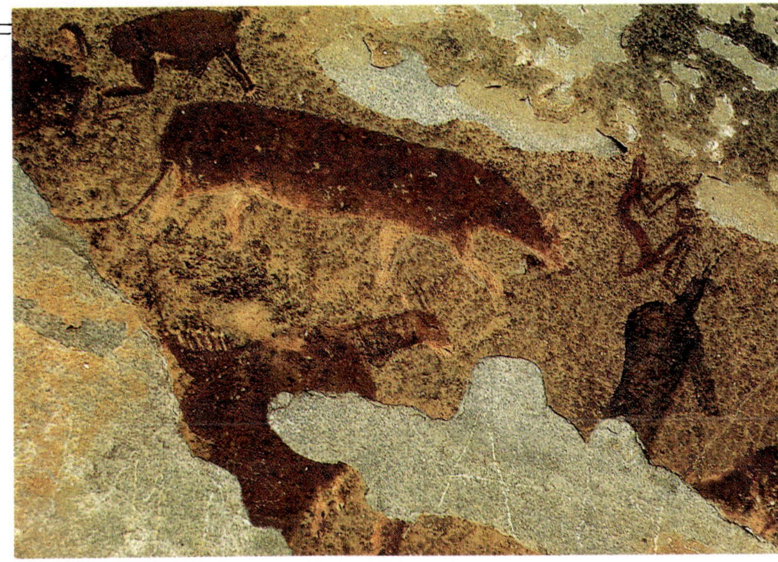

Figure 113

LIGHTNING

According to Bushman folklore the lightning is sent by the rain (Bleek & Lloyd, 1911, 393-7) especially to carry off girls — who become stars — 'like a flower which grows in the water'. Water lilies certainly resemble stars. Frobenius was of the opinion that several Bushman paintings recorded lightning (Frobenius, 1931) and it would seem likely that such a powerful force would have attracted the attention of artists, especially when painting scenes that involved rain. The zigzag lines connecting the groups of eland in the Clocolan and Bethlehem districts (Figures 94, 95) may well depict lightning as may several of those at Beersheba (Figure 87). The zigzags at the top of the 'sacrifice' painting dealt with in chapter 15 may also be derived from the lightning model.

In the Underberg District the fragmentary remains of what must have been an outstanding masterpiece include a seated therianthrope with a tail and raised hands. From the hands lines extend that could represent lightning (Figure 113).

The folklore tells of how *kagara* and *!haunu*, the first a bird and the second a rain sorcerer fought each other with lightning; then *kagara* took his little sister away from *!haunu* to whom she was apparently married. The narrator said that when the clouds were thick and black and 'resembled a mountain' and there was lightning, his grandmother used to say: 'It is *kagara* with *!haunu*'. *Kagara* was the winner of the fight, according to the story, and his final success was the result of striking *!haunu* with 'black lightning' which, 'whisked him up (and carried him to a little distance)'.

There are many other references to lightning in the folklore so it seems highly likely that it would also have been represented among paintings concerned with rain — and appropriately by a zigzag line.

Above: 'Lightning-man'? in top right corner. Underberg, Natal.

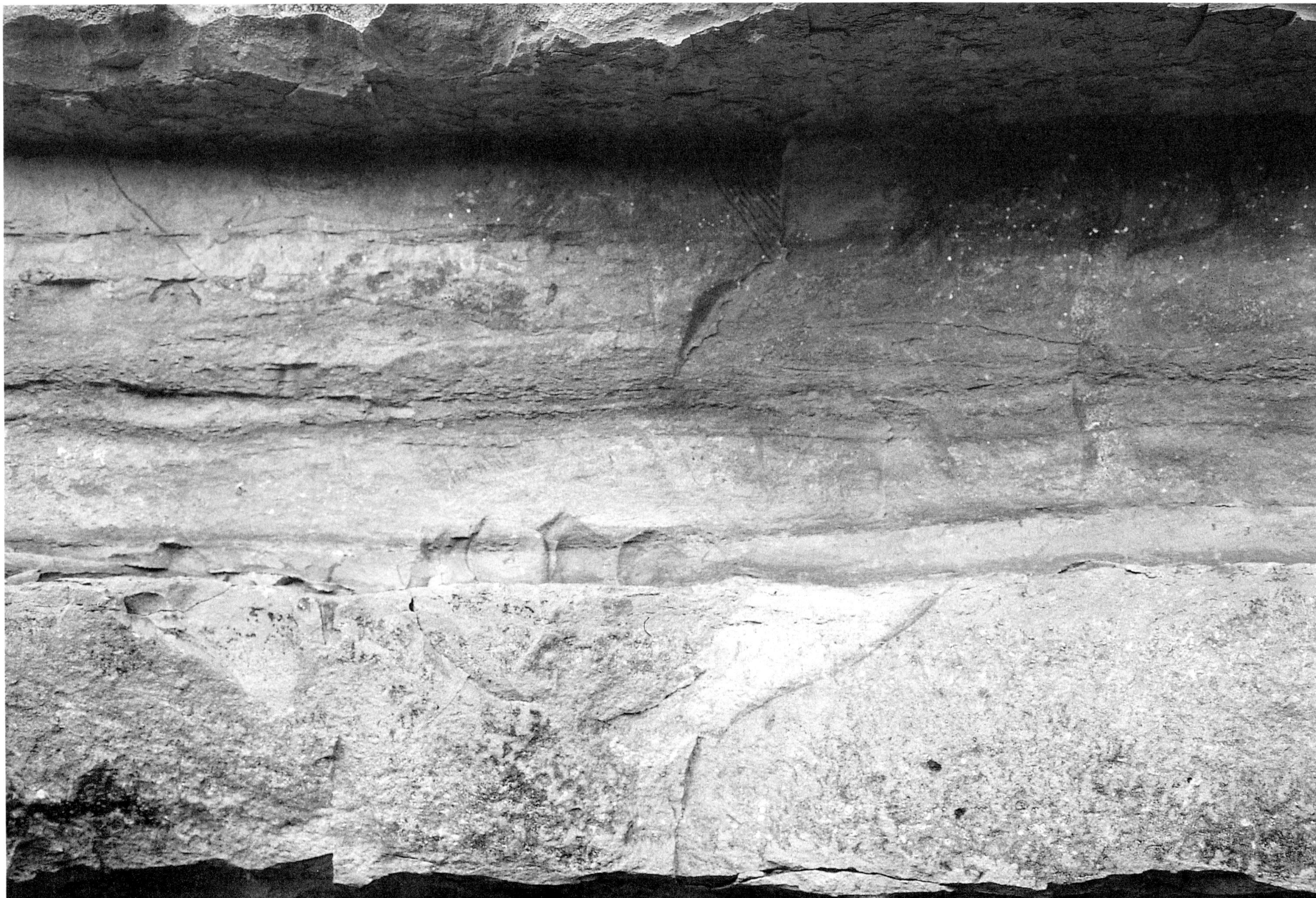

Figure 114: The forequarters of a large rain feline which appears to be emitting clouds from its open mouth. Rhodes.

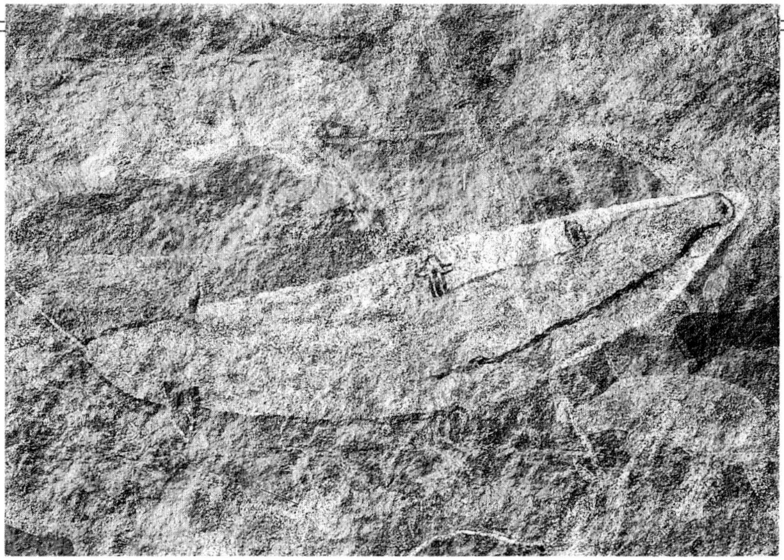

Figure 115

ENIGMAS

In the Rouxville District at the Cave of the White Rhinoceros (Stow, 1905, 33) (Figure 116) is a painting that was copied by Stow and reproduced in Bleek & Stow, (1930, plate 41) (Figure 117). It was captioned as a rain animal and in the Stow copy it could be regarded as the head of a large animal with pointed ears sloping backwards. My photograph, taken almost one hundred years after Stow made his copy, does not convey quite the same impression — but if not a rain animal, what is it? The object immediately above the 'rain animal' and obviously part of the same picture is particularly puzzling. It is clearly defined and is not eroded or vandalised in any way. It is apparently the same now as when Stow copied it more than a hundred years ago. But again, what is it? Does it have any relationship to rain? It certainly provides a 'cut-off' to the animal head and neck above it in a manner similar to the technique used in, for example, figures 99 and 101, and this whole scene is in close proximity to the rain animal reproduced in Figure 46. The lower figure also includes vertical shower lines.

A search for comparative paintings has led to one that was first recorded by Walter Battiss and Murray Schoonraad in the Maclear District and reported in the *Sunday Times* in the early 1960s (Figure 115). They suggested that it might represent a ship — even the ill-fated *Grosvenor*. The search also produces a more concave version (Figure 120) which is immediately above the rain animal scene depicted in Figure 66.

Close to the 'rain snake emergent' is a strange depiction at a shelter in the Clocolan district (Figure 118). The slug-like animal on the left of Figure 89 also seems to be in the same category, and mingled with an extensive frieze of eland and elephant at Kalkoenkrans near Aliwal North is a somewhat fatter version with a beak-like head (Figure 119). There is a porpoise-like

Above: Enigma variation. Maclear.

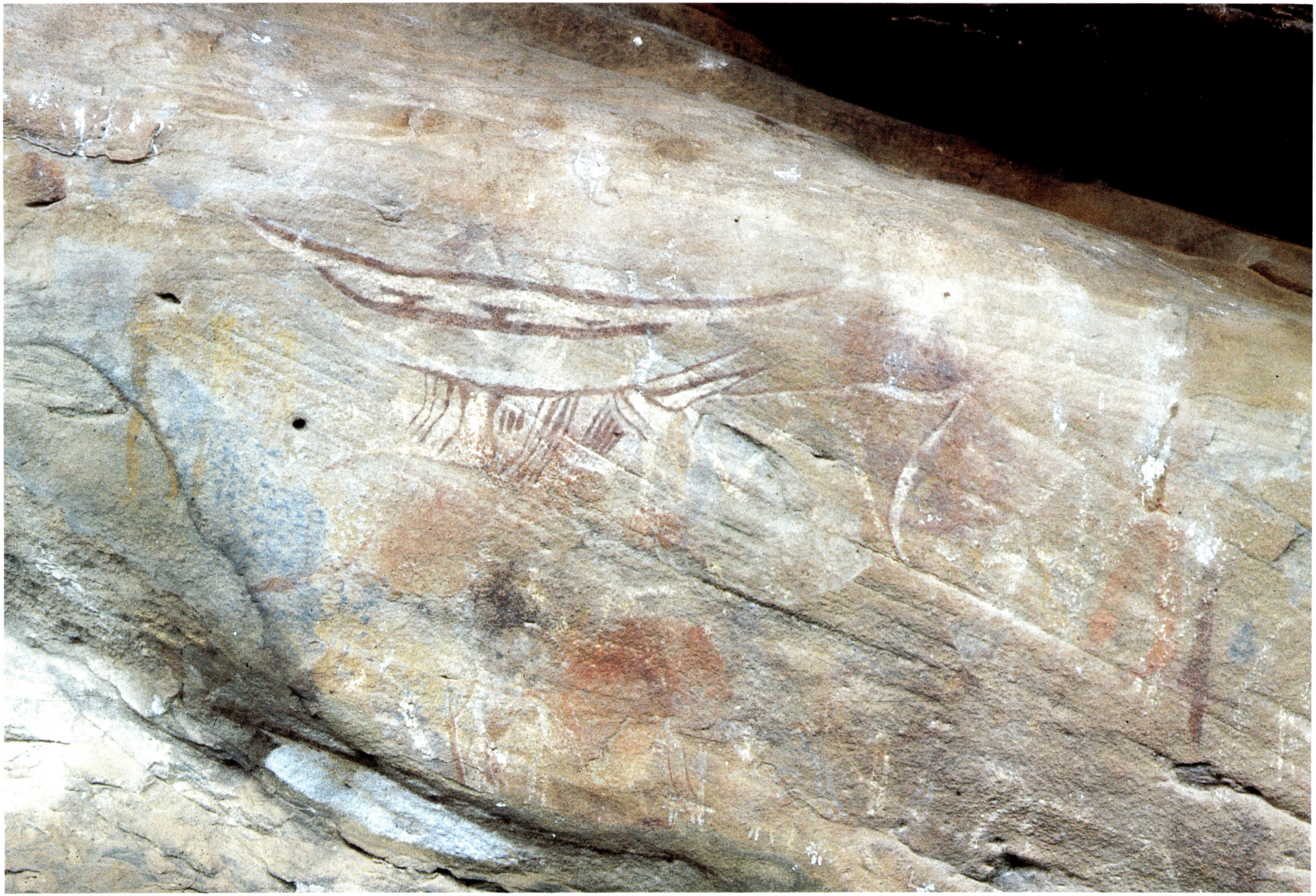

Figure 116: Rain animal with 'enigma'. Rouxville. Photographed in 1977.
See opposite for Stow's copy.

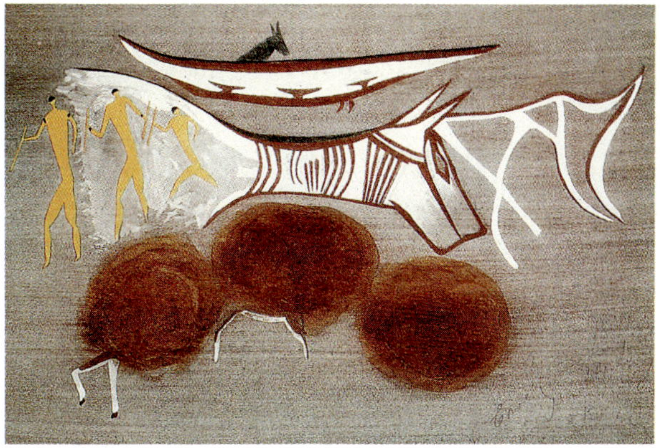

Figure 117

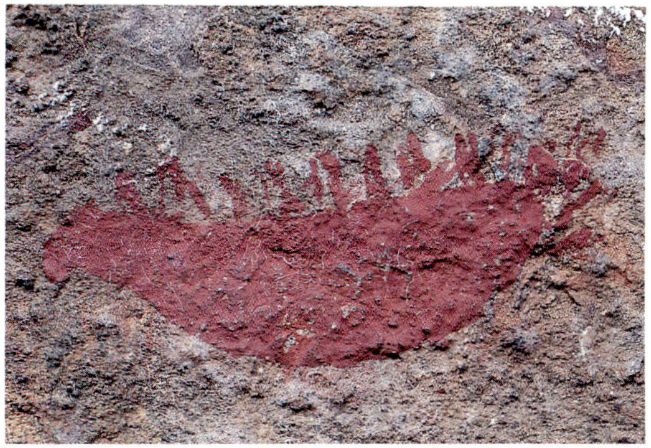

Figure 118

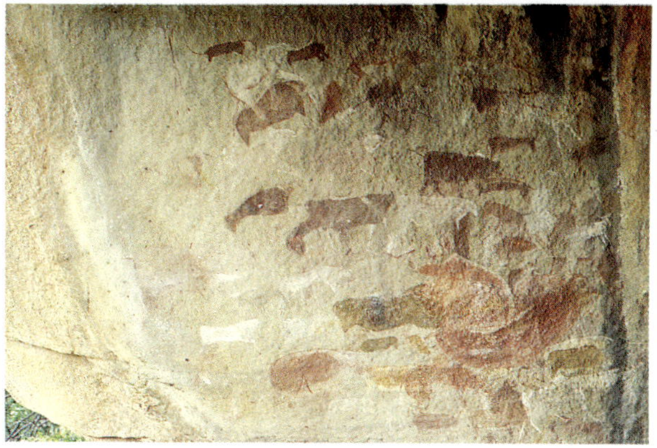

Figure 119

Figure 117: Stow's copy (1870) of a rain animal with 'enigma'. Rouxville.

Figure 118: Enigma variation. Clocolan.

Figure 119: Slug-like creature with a bird-like head at lower right mixed in with a herd of eland and elephants. Aliwal North.

Figure 120: Enigma variation. Transkei.

Figure 121: Porpoise-like creature which is a detail of a frieze in Wedza, Zimbabwe.

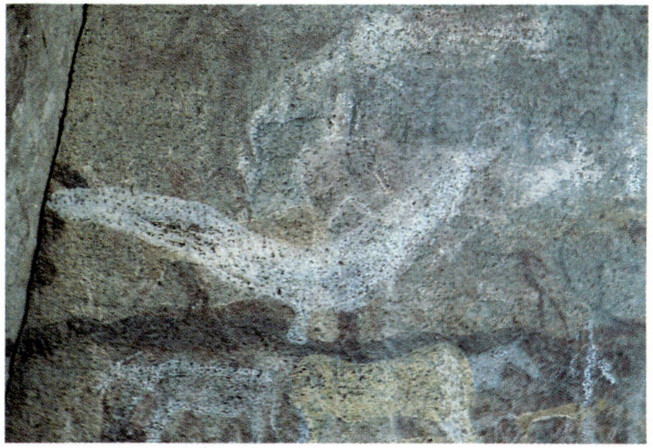

Figure 120

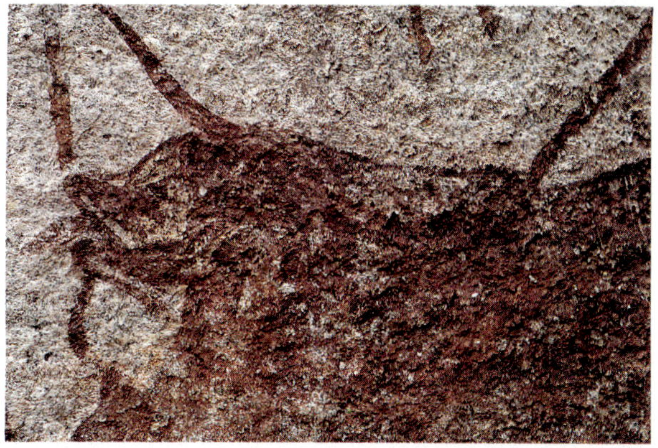

Figure 121

example at Wedza in Zimbabwe associated with a large rain animal and a river (Figures 122 and 121) and another is illustrated in Figure 89.

In the Smithfield district there is a painting of a rather amorphous 'slug' surrounded by bending and leaping figures involved in a veritable orgy of dancing — presumably to conjure up the required rain or to mark the actual arrival of rain. Modern ethnographers have recorded the excitement and impromptu dancing that takes place on the arrival of rain in the Kalahari.

Certainly the painting of the slug-like animal, and similar ones, all seem to be associated with other paintings indicative of rain or rainmaking but their exact meaning remains enigmatic although they may well be regarded as cloud-like. Perhaps they were inspired by 'mushroom cloud' formations.

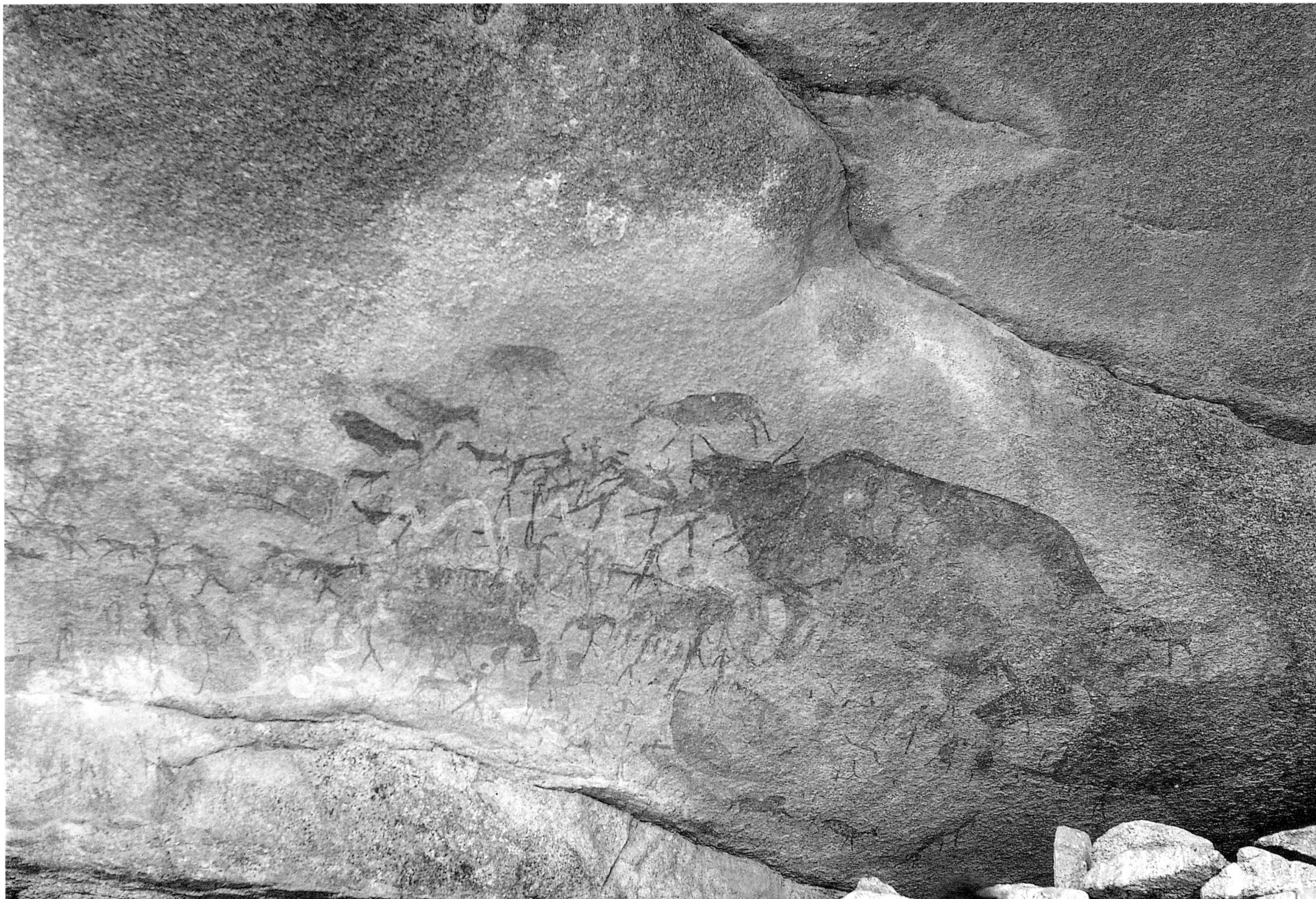

Figure 122: Rain animal, river in white and many other paintings. Wedza, Zimbabwe.

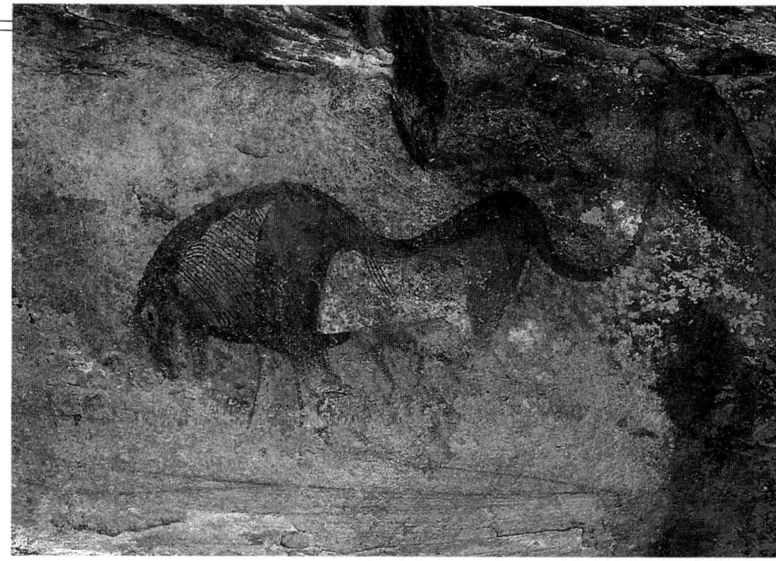

Figure 123

RAIN LIONS

The fear of an 'angry' rain is a recurrent theme in Bushman folklore. An account of a rain sorcerer and his friends, 'the old men' included an entreaty to the rain in the words:

> O beast-of-prey, let us now return home, for I have been dreaming about rain falling, the rain I dreamt about was an angry rain, for it lightened not a little. For I dreamt that it lightened very strongly at us, so that we hid from the thunderstorm. [Bleek, 1933b, 389].

The fact that the rain was addressed as a beast of prey is significant in understanding a painting from the Clocolan district (Figure 124) in which a white therianthropic figure with wings is seated in the top left-hand corner superimposed on three red paintings of fish. The winged figure is clearly an important part of a scene which includes several eland and a lion-like animal, all of which are linked with a white line. White is a colour which is used by a number of 'Bantu-speaking' tribes at times of great significance, e.g. initiation ceremonies, when in close contact with ancestral spirits (e.g. Broster, 1981), and many of the Bushman paintings in white depict spirit-like creatures.

The animal in the centre of the Clocolan painting has an ethereal air and the stooping figure on the extreme left, partly white and partly purple, is also likely to be a non-literal representation. The lion-like creature has a tusk protruding from its jaws in a manner similar to that in a scene which depicts lion-like creatures in the Fouriesburg district (Figure 125). That particular scene seems to suggest a further example of the leading/driving of the

Above: Rain lion? Removed from a site near Wepener to the National Museum, Bloemfontein. Photographed while still in situ, *1970.*

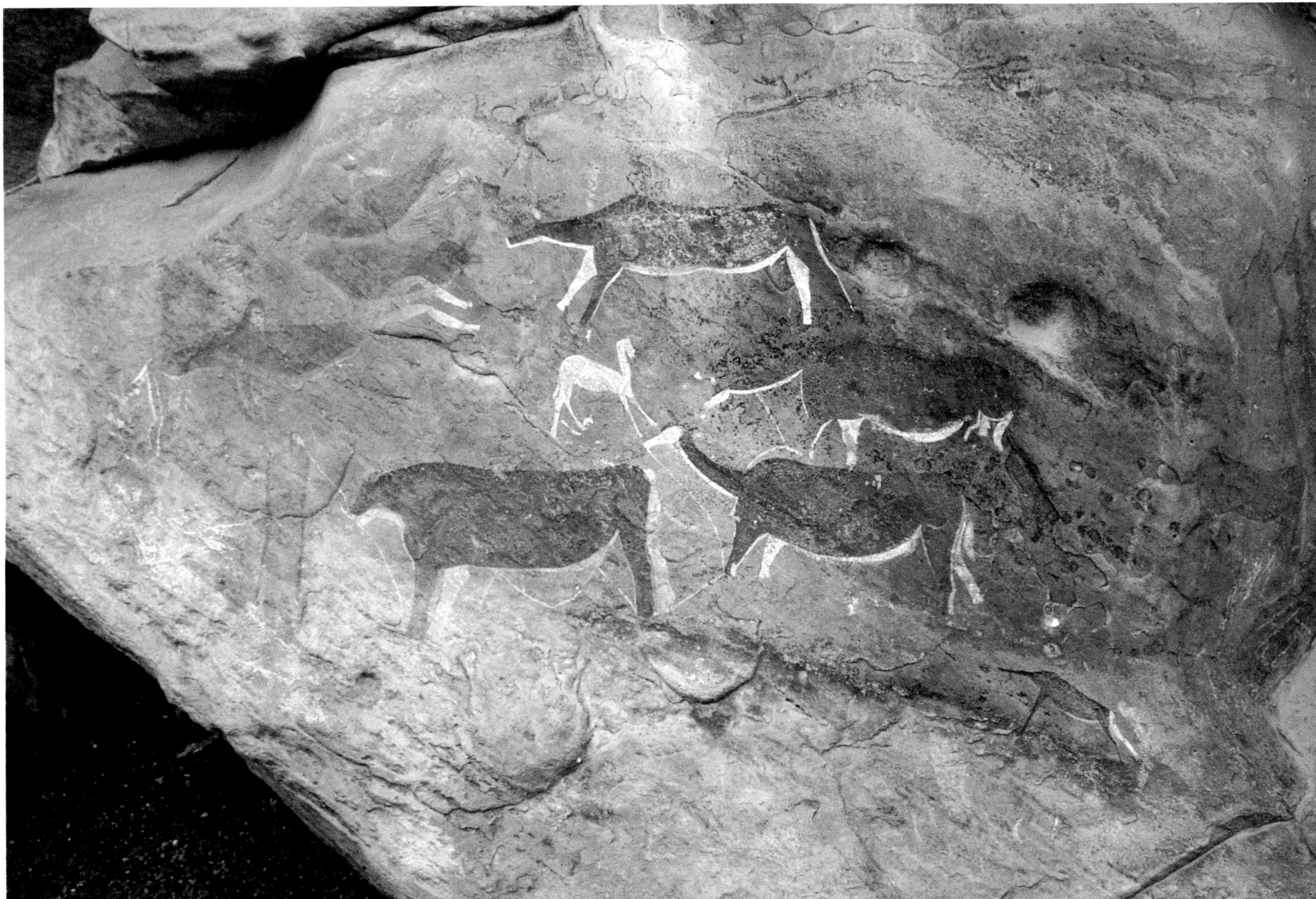

Figure 124: A white winged therianthrope is seated on top of three red fish in the top left corner of this highly imaginative painting of rain animals. One of the animals is feline but has a tusk. Clocolan.

Figure 125:

Two felines with tusks being driven. Fouriesburg.

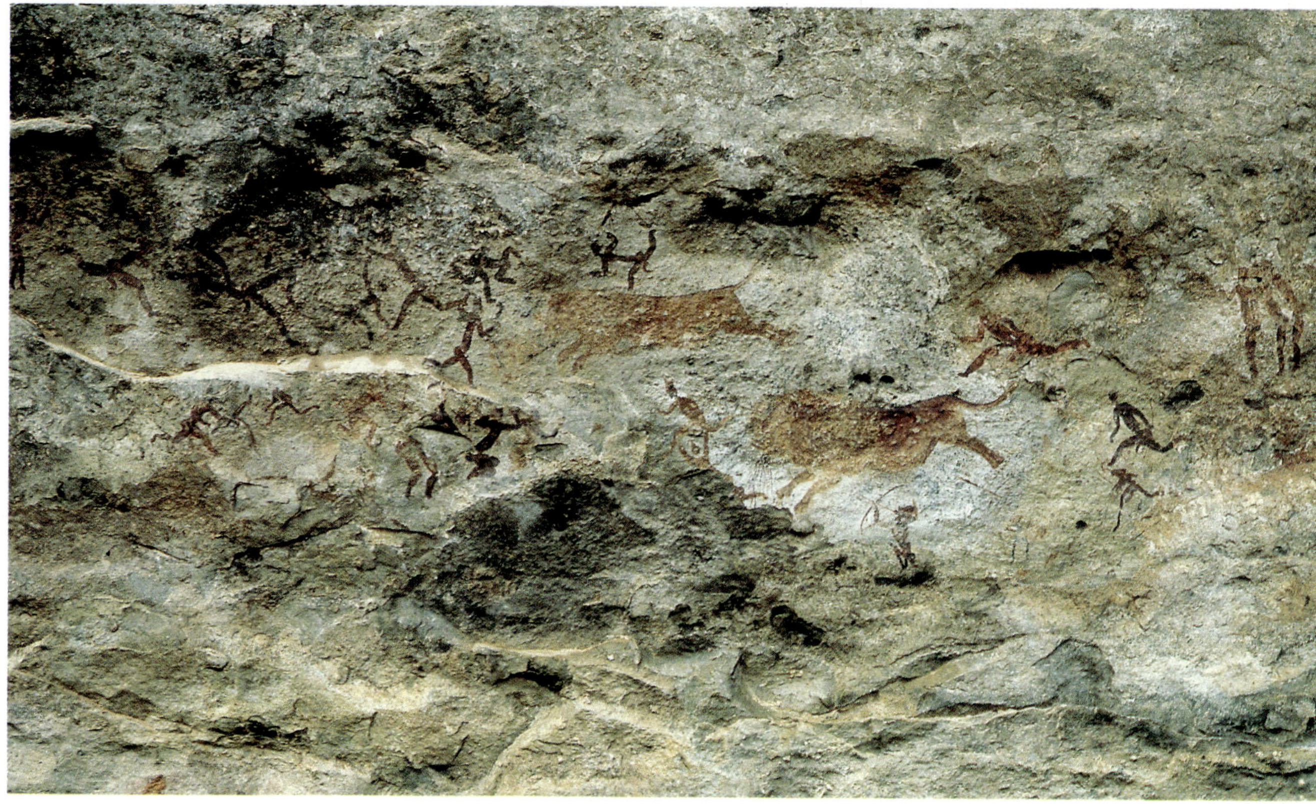

Figure 125

rain animal but the sorcery has gone wrong and an angry 'beast-of-prey' rain has been aroused in the form of two lions. The quiescent nature of the Clocolan painting establishes various patterns of rain-making entreaties in the form of the herd of eland with their spiritual representation, the white animal, and the lion, but there is no indication as to which kind of rain - gentle or angry - will eventually appear.

This painting was first reproduced in an article on creatures with tusks (Woodhouse, 1987) and the Fouriesburg scene appeared in *Art on the Rocks* (Lee & Woodhouse, 1970, Plate 52) although at that time we considered it to be a straightforward hunting scene.

Once the likelihood is accepted of there being a rain lion, one of the most striking representations would be the painting removed from the Wepener district to the National Museum in Bloemfontein - a step taken to avoid the possibility of its being inundated by the water of the P K le Roux Dam. That particular lion (Figure 123) has a 'mane' consisting of very straight black lines which could be construed to symbolise rain showers. A short distance away from it, there is a rather more docile-looking rain animal whose back legs have hooves but which are otherwise human in appearance. It approaches two therianthropes who, regrettably, merge into a stained rock face. These creatures heighten the general impression of a rain-making scene, but much of it had already been weathered away at the time of the removal of the black-maned lion (Figure 126).

There is a huge feline-type creature painted in the Rhodes district which appears to be exhaling fluffy cirrus clouds. One is reminded of the fact that the G/wi Bushmen personified rain as a giant leopard (Silbebauer, 1965, 101) (Figure 114).

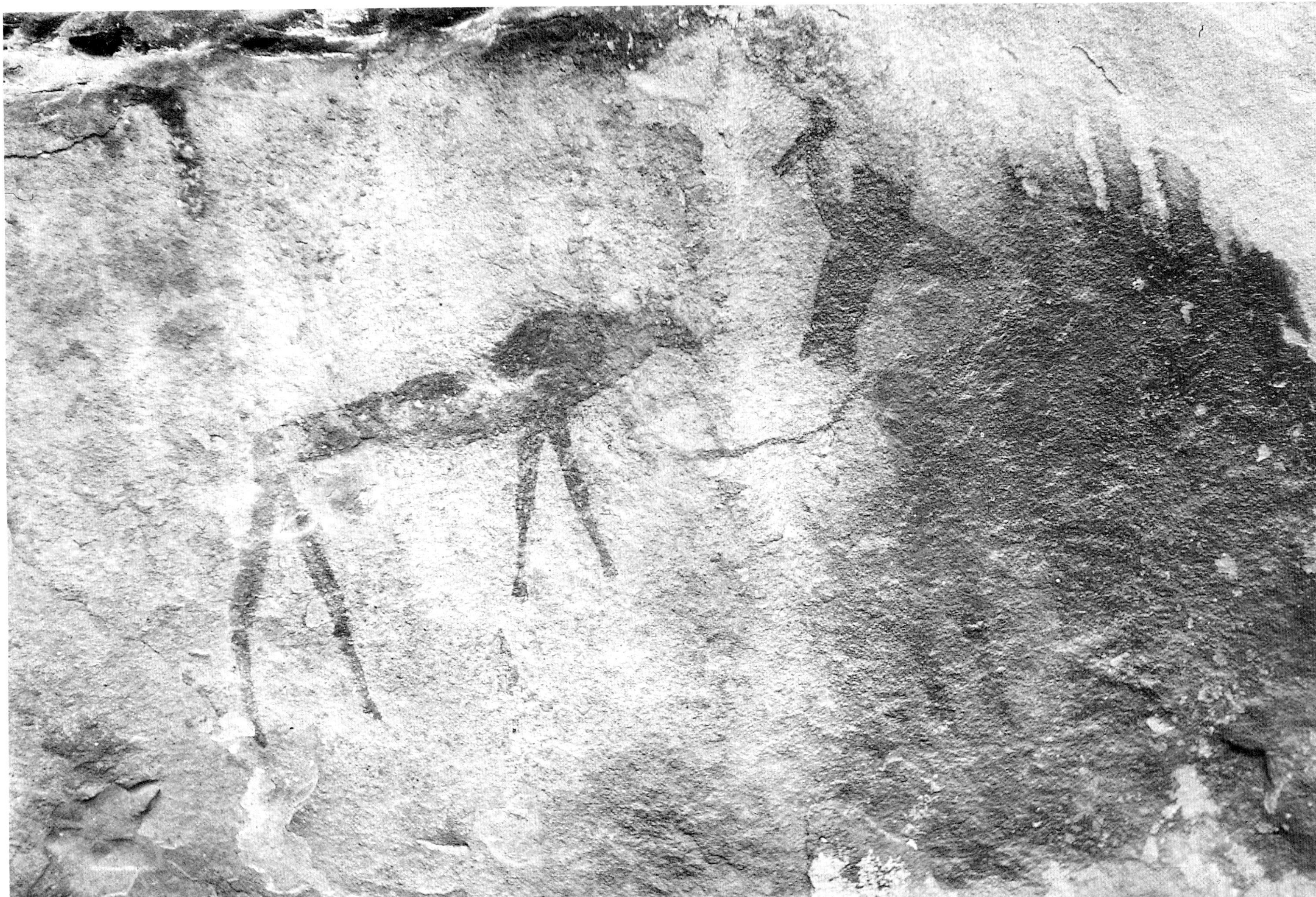

Figure 126: Therianthropic creatures at a site near Wepener.

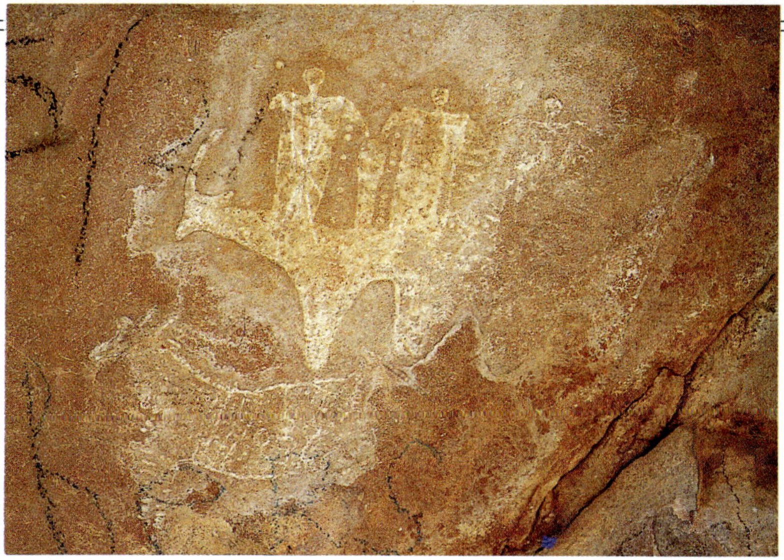

Figure 127

'Dead People' and the Rain

Hankasso, one of Dr Bleek's Bushman informants said that the old men speak to the dead men who are with the rain in the following words:

O gallopers,
O gallopers,
Do you not know me?
You do not seem to know my hut [Bleek, 1933, 304].

This was a complaint for being overlooked because 'Dead people who come out of the ground are those of whom my parents used to say, that they rode the rain, because the thongs with which they held it were like the horse's reins, they bound the rain. Thus they rode the rain because they owned it'. This concept may be relevant to a painting in the Northern Transvaal (Figure 127) which I have only seen in a photograph sent to me by enthusiast Fanie Venter.

People wearing karosses and grouped together in amorphous patterns like clouds, with white faces and with 'rain legs' falling from them, together with a vertical rain snake like that in Figure 17 and a variety of rain animals, are blended into an amazing series of compositions (Figures 128-130 and Figure on pviii) in the Clanwilliam district. Part of the painting at this site was first published by Jalmar and Ione Rudner (Rudner, J & I, 1970, plate 40) who hesitated to relate it to any particular piece of folklore but drew attention to rain-making activities in general and the fact that, as recorded by

Above: People who rode the rain? Photograph kindly sent to me by enthusiast Fanie Venter in the Northern Transvaal.

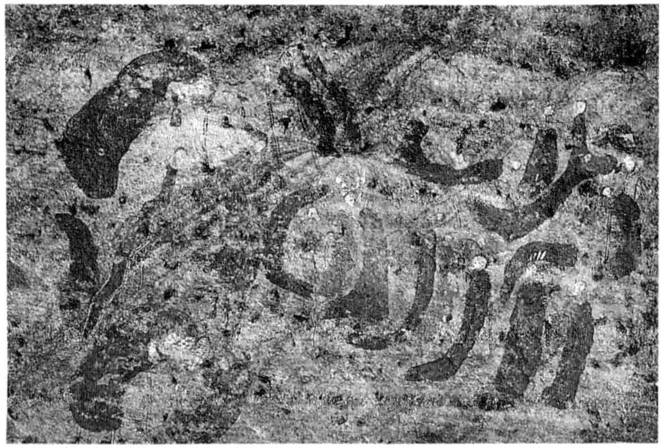

Figure 128: Personification of rain clouds? Clanwilliam.

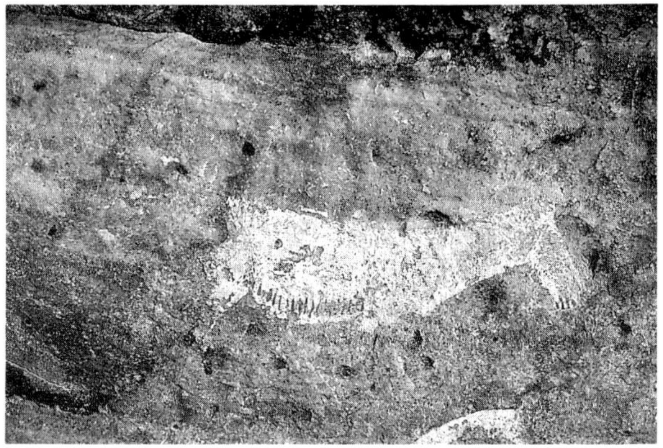

Figure 129: Rain animal, part of the same scene in the Clanwilliam painting.

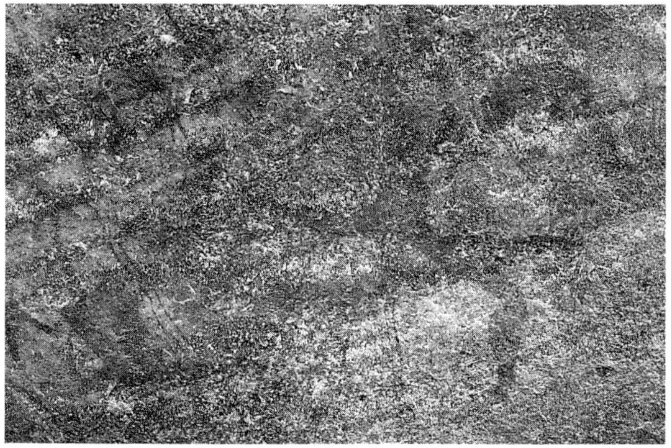

Figure 130: Source figure? Part of the scene in the Clanwilliam painting.

Sonia Cole, various people of Tanzania still carried out such ceremonies by sacrificing animals at rock-art sites and painting over the paintings with the fat of the slaughtered beast (Cole, 1954).

Although the cloud-like people in the Clanwilliam painting are not 'riding' the rain or holding it with thongs they evoke similar feelings to those aroused by the notion of dead people in the clouds. They exude little 'showers', in the form of what Hankasso called 'a little rain leg', from various parts of the body. At least two of the animals with them are very similar to the zebra-like creatures to the left of the prototype rain animal considered in chapter 2 which provides a link-up over a considerable geographical distance.

A further link-up at the same site is provided by a human 'source figure' (Figure 130) which has two lines from between the legs like those recorded by Mrs Goodall from the Mtoko district mentioned earlier.

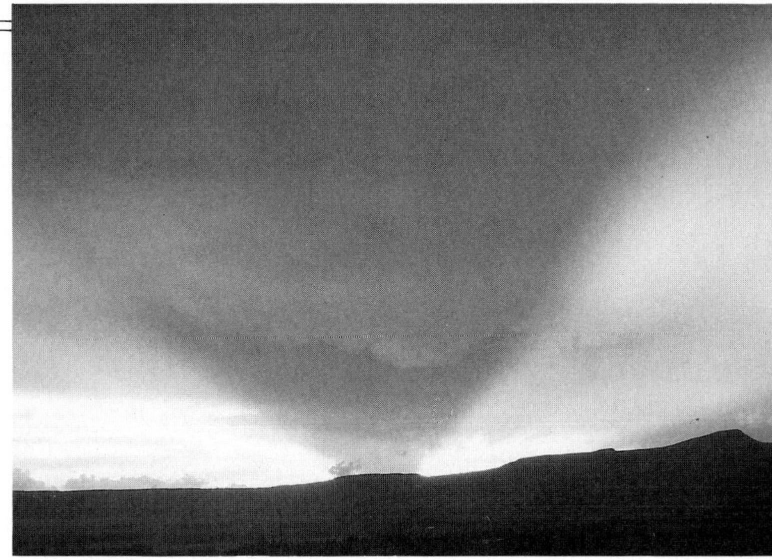

Figure 131

ROUND-UP

The review of rain animals and rainmaking began with a prototype in the Fouriesburg district of the Orange Free State and ended with a very sophisticated example in the Clanwilliam district of the western Cape Province. In-between, the scope has ranged far and wide over the sub-continent, from Mtoko in eastern Zimbabwe to the Matopos near Bulawayo; from the Sani Pass in Natal to the mountains of the Uitenhage district in the eastern Cape and to the Herschel district where the Cape, the Orange Free State, Lesotho and the Transkei come together to provide a 'heartland' for these fascinating examples of art on the rocks. Appropriately, there are also examples in the Waterberg of the Transvaal where exploration for rock art is still in its infancy, but where there are some accessible examples in the Lapalala Nature Reserve.

After considering so many examples of variations on a theme — even if some remain enigmatic - one cannot help concluding that a large part of the lives of the artists was concerned with rain, what sent it and what withheld it; how to attract a soft, life-giving 'she-rain' and how to dispel a fierce, destructive, 'he rain'. In some cases perhaps the artists were also the rain sorcerers, medicine-men or shamans, but perhaps not. In many instances, they would have applied personal interpretations to their art. I have looked at records and communications - but no two representations were the same. They were, above all, testimony to the unique nature of every individual artist and his personal version of the 'world view' of the group of which he was a part - albeit an exceptional one.

Perhaps this round-up of creatures of the rain is best illustrated by a specimen (Figure 132) from the Wodehouse District near Dordrecht, which might be regarded as equally important to the prototype animal near Bethlehem. It was first recorded by Helen Tongue and published in 1909. I photographed it in 1990. It does not appear to have deteriorated in the intervening eighty

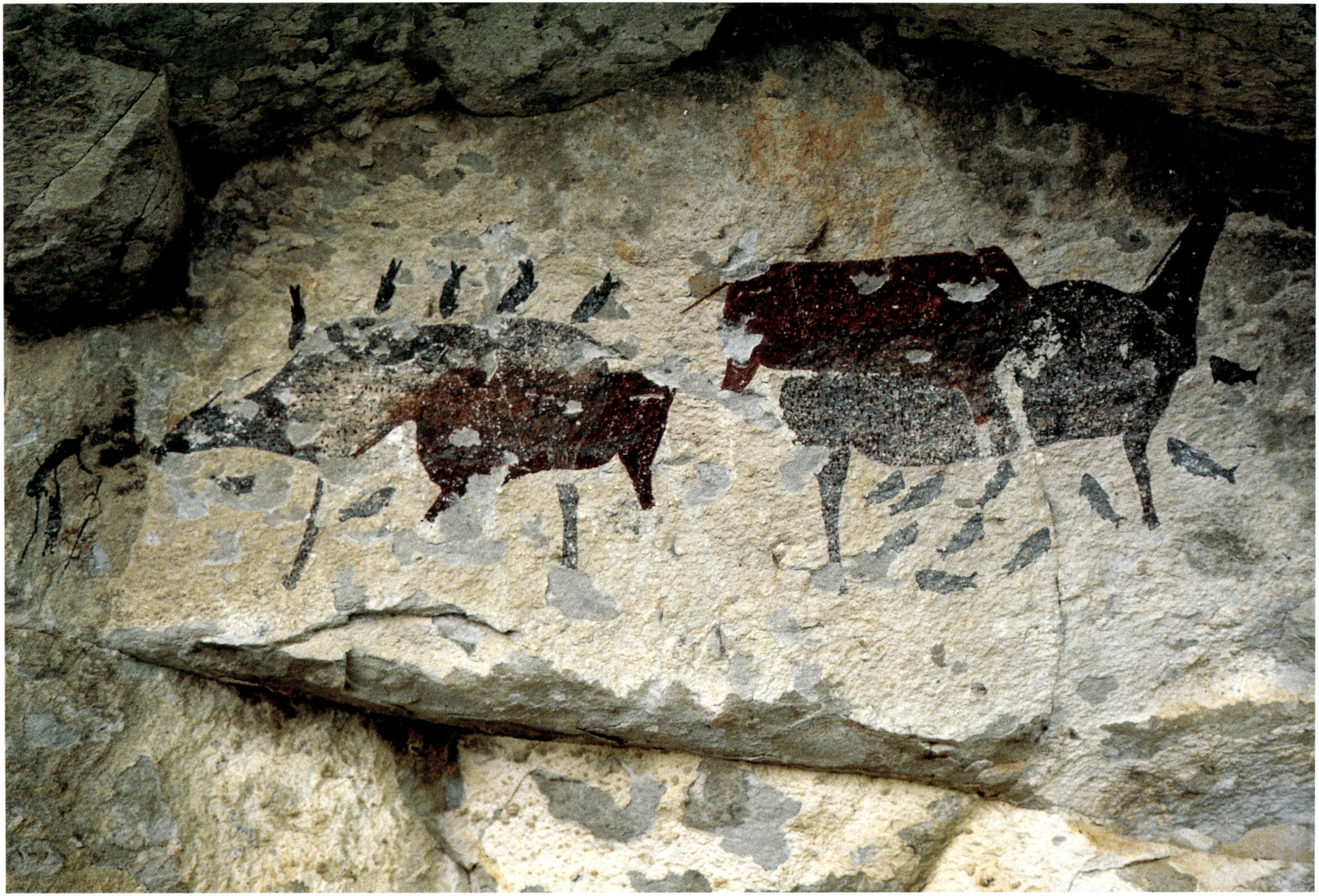

Figure 132: Originally recorded by Helen Tongue in the early years of this century. She visited sites in a pony and trap. Note old man holding fish — bounty from the rain animals — on left. Wodehouse.

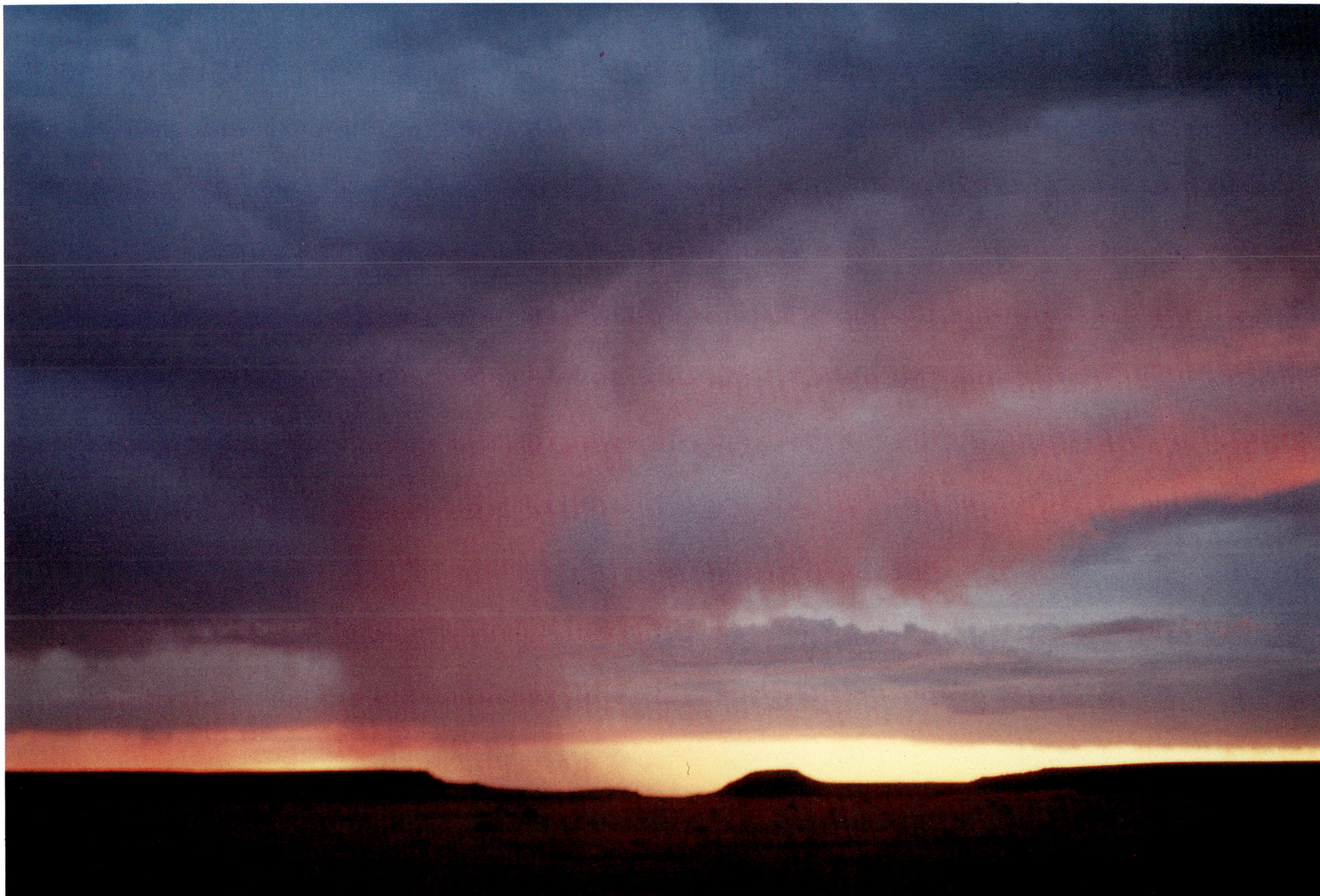

Figure 133: The rain still has legs.

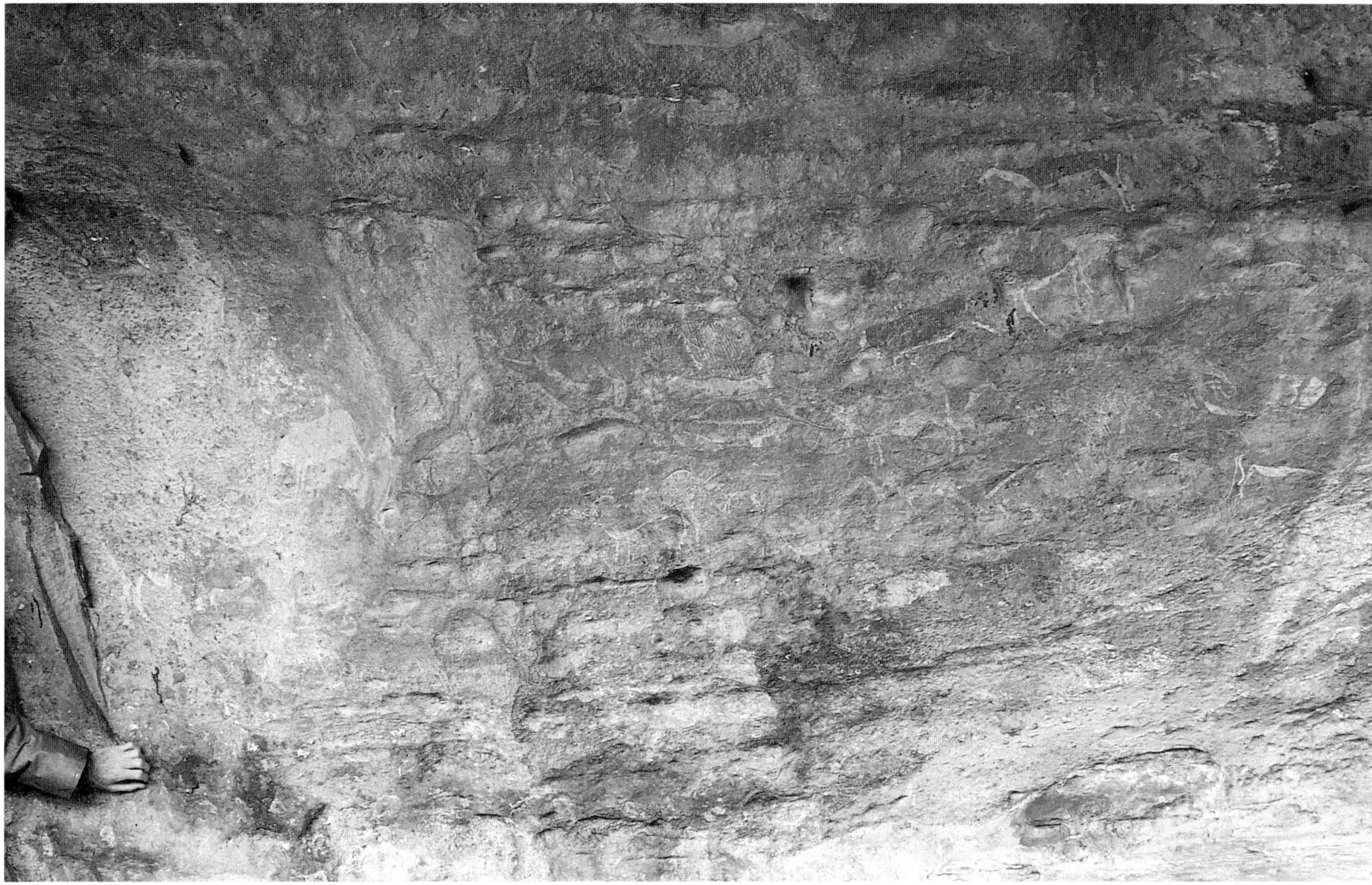

Figure 134: Frieze of eland being driven over a cliff in a rainstorm by a man carrying a staff and with the tusked head of a rhinoceros or warthog — a therianthropic creature — presumably the medicine man of the rain.

years. Like the Bethlehem painting, it consists of two rain animals. The originals were black but later versions have been superimposed in red. The black animals are surrounded by fish. One animal stretches its neck to browse while the other lowers its head to graze. To the left a feeble old man carries a fish. All these actions are made possible by the rain and its creatures.

Rain, water, rainmaking, rain animals and rock art: Significant steps in the study

1867-1882	G W Stow made copies of rock paintings.
1874	Copies sent to Dr W H I Bleek and Lucy Lloyd. Referred to Bushman informants and comments noted.
1886	Magistrate V H Stanford took statement from Silayi, rain helped in cattle rustling. Noted that Bushmen were rainmakers to Pondomise.
	Magistrate J M Orpen recorded conversations with Bushman Qing, in Cape Monthly Magazine. Reference to 'snake' and to 'leading the rain animal'.
1908	Magistrate Hook — in book *With Sword and Statute* referred to only Bushmen known in Kaffirland practising as rainmakers in krantz of Buffalo River.
1911	*Specimens of Bushman Folklore*. Bleek and Lloyd. References to medicine men of the rain, 'he rain' and 'she rain' and fear of heavy storms.
1928	Many of Stow's sites revisited by Dorothea Bleek. M C Burkitt, archaeologist from Cambridge visited Southern Africa and published *South Africa's Past in Stone and Paint*. Description of rainmaking at Domboshawa near Salisbury (Harare).
1930	Publication by Dorothea Bleek of Stow's copies with identification of rain animals and comments by Bushmen from Northern Cape.
1976	Vinnicombe, P. *People of the Eland* pp 323-345.
1979	Schmidt, S. The rain bull of the SA Bushmen. *African Studies* 38.2.
1981	Lewis-Williams J D Chapt. 8 The rain and the eland in *Believing and Seeing* 103-116.
1986	Jolly P. A first generation descendant of the Transkei San. Informant's sister was rainmaker to the Pondomise. *South African Archaeological Bulletin*. 143. 6-9.
1988	Woodhouse H C. *Motivation and the rain animal in the rock art of Southern Africa*. Paper delivered to a symposium on rock art at Rand Afrikaans University.
1989	Above paper published in *SA Journal of Art History* 4 (3 and 4).
1990	Prins F E. Southern-Bushman descendants in the Transkei *SA Journal of Ethnology* 13.
1992	Woodhouse H C. *The Rain and Its Creatures* — William Waterman Publications.

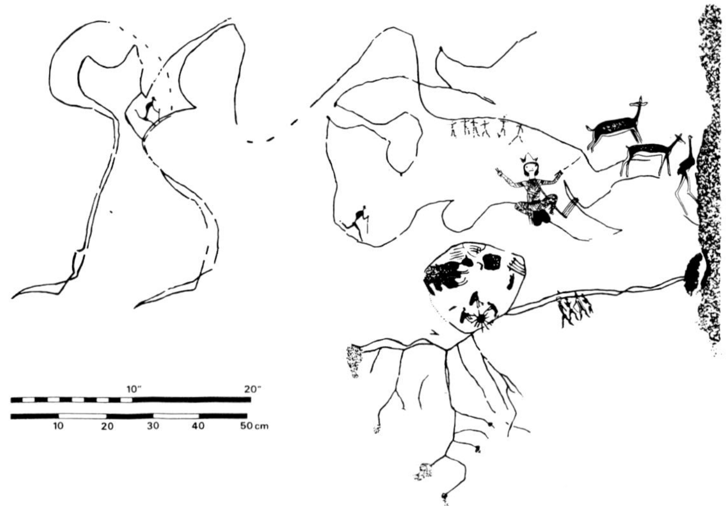

REFERENCES

ANDERSON, W. 1907. *Third and final report of the geological survey of Natal and Zululand.* London: West, Newman.

BAARD, E. 1970. 'The "white lady" of Kalkoenkrans'. *South African Journal of Science.* 66.5: 151-154.

BATTISS, W W. 1948. *The Artists of the rocks.* Pretoria: Red Fawn Press.

BLEEK, D F (ed). 1923. *The mantis and his friends.* Cape Town: Maskew Miller.

BLEEK, D F. 1928. *Bushmen of central Angola.* Bantu Studies (3): 2.105/125.

BLEEK, D F. 1933a. 'Beliefs and customs of the !Xam Bushmen. Part V The rain'. *Bantu Studies.* V11.3.

BLEEK, D F. 1933b. 'Beliefs and customs of the !Xam Bushmen. Part V1 Rainmaking'. *Bantu Studies.* V11.4.

BLEEK, W H I & **LLOYD,** L C. 1911. *Specimens of Bushman folklore.* London: George Allen.

BROSTER, J A. 1981. *Amagqirha.* Goodwood: Via Afrika.

BURKITT, M C. 1928. *South Africa's past in stone and paint.* London: Cambridge University Press. 119-20.

COLE, S. 1954. *Prehistory of East Africa.* Harmondsworth: Penguin.

COOKE, C K. 1963. 'Report on excavations at Pomongwe and Tshangula caves, Matopo Hills, Southern Rhodesia'. *South African Archaeological Bulletin.* 18: 73-151.

DORNAN, S S. 1923. *Rainmaking in South Africa.* Bantu Studies. 3: 185-95.

DOWSON, T A & **HOLLIDAY,** A. 1989. 'Zigzags and eland : an interpretation of an idiosyncratic combination'. *South African Archaeological Bulletin.* 44: 46-48.

FOSBROOKE, H A. & others. 1950. 'Tanganyika rock paintings'. *Tanganyika Notes and Records.* 29.

FROBENIUS, L. 1931. *Madsimu Dsangara.* Graz: Akademische Druck-u Verlagsanstalt.

GOODALL, E. 1949. 'Notes on certain human representations in Rhodesian rock art'. *Transactions of the Rhodesian Scientific Association.* 42: 69-72.

HARDING, J R. 1950. A remarkable rock painting in the Bethlehem district, Orange Free State. *South African Archaeological Bulletin.* 5: 51-54.

HOLM, E. 1961. in Bandi ed. 'Rock Art of South Africa'. *Art of the world, the Stone Age.* London: Methuen.

HOOK, D B. 1908. *With sword and statute.* London: Greaves, Pass & Co.

HOW, M W. 1962. *The mountain Bushmen of Basutoland.* Pretoria: Van Schaik.

HUFFMAN, T N. 1983. 'The trance hypothesis and the rock art of Zimbabwe in new approaches to southern African rock art'. Goodwin series 4. *South African Archaeological Society.*

JOHNSON, R **TOWNLEY**, 1979. *Major rock paintings of Southern Africa.* Cape Town: David Philip.
KIRBY, P R. 1968. *The musical instruments of the native races of South Africa* 2nd Edition. Johannesburg: Witwatersrand University Press.
KNUFFEL, W E. 1973. *The construction of the Bantu grass hut.* Graz: Akademische Druck-u Verlagsanstalt.
KRIGE, E C B. 1936. *The social system of the Zulus.* Pietermaritzburg: Shuter & Shooter.
KRIGE, E J & J D. 1980. *The realm of a rain queen.* Cape Town: Juta.
LARSON, T J. 1966. 'The significance of rainmaking for the Mbukushu'. *African Studies.* 25: 23-36.
LEE, D N. 1968. 'Bushman folklore and rock paintings'. *South African Journal of Science.* 68: 195-99.
LEE, D N & **WOODHOUSE**, H C. 1964. 'Rock paintings of 'flying buck'. *South African Archaeological Bulletin.* 19: 71-74.
LEE, D N & **WOODHOUSE**, H C. 1970. *Art on the rocks of Southern Africa.* Cape Town: Purnell. 131.
LEWIS-WILLIAMS, J D. 1977a. 'Ezeljagdspoort revisited: new light on an enigmatic rock painting'. *South African Archaeological Bulletin.* 32: 165-169.
LEWIS-WILLIAMS, J D. 1977b. 'Led by the nose: observations on the supposed use of southern San rock art in rainmaking rituals'. *African Studies.* 36.2: 159-159.
LEWIS-WILLIAMS, J D. 1981a. *Believing and seeing.* London: Academic Press.
LEWIS-WILLIAMS, J D. 1981b. 'The thin red line : southern San notions and rock paintings of supernatural potency'. *South African Archaeological Bulletin.* 36: 5-13.
LEWIS-WILLIAMS, J D. 1983. 'Introductory essay: science and rock art' in Lewis-Williams, J D (ed). 'New approaches to southern African rock art'. *South African Archaeological Society.* Goodwin Series 4.
LEWIS-WILLIAMS, J D & **DOWSON**, T A. 1988. 'The signs of all times'. *Current Anthropology.* 29: 2, 201-45.
LEWIS-WILLIAMS, J D & **DOWSON**, T A. 1990. 'Through the veil: San rock paintings and the rockface'. *South African Archaeological Bulletin.* 45: 5-16.
MAGGS, T M O'C. 1979. 'Naming of archaeological sites'. *South African Archaeological Bulletin.* 34: 110.
MAGGS, T M O'C. & **SEALY**, J in Lewis-Williams, J D. (ed) 1983. 'Elephants in boxes'. *South African Archaeological Bulletin.* Goodwin Series 4.
MARSHALL, L. 1957. N!ow. *Africa.* 24: 232-40.
MARSHALL, L. 1962. '!Kung bushman religious beliefs'. *Africa.* 32: 234.
MARSHALL, L. 1969. 'The medicine dance of the !Kung bushman'. *Africa.* 39.
ORPEN, J M. 1874. 'A glimpse into the mythology of the Maluti Bushmen'. *Cape Monthly Magazine.* 9.
PAGER, H. 1971. *Ndedema.* Graz: Akademische Druck-u. Verlagsanstalt.
PAGER, H. 1975. *Stone age myth and magic.* Graz: Akademische Druck-u. Verlagsanstalt.
RAUTENBACH, G B. 1967. 'A late stone age shelter in the Clarens district, Orange Free State'. *South African Archaeological Bulletin.* 21: 183-86.
REYNOLDS, B. 1968. *The material culture of the peoples of the Gwembe valley.* New York: Frederick A Praeger.
RITTER, E A. 1955. *Shaka Zulu.* London: Longmans.
ROSENTHAL, E & **GOODWIN**, A J H. 1953. *Cave Artists of South Africa.* Cape Town: Balkema.
RUDNER, J & I. 1970. *The hunter and his art.* Cape Town: Struik.
SCHAPERA, I. 1971. *Rainmaking rites of the Tswana tribes.* Cambridge: African Studies Centre. 27, 35.
SCHMIDT, S. 1979. 'The rain bull of the South African Bushmen'. *African Studies.* 38: 201-24.
SILBERBAUER 1965. 'Bushman survey report'. Bechuanaland Government: Gaberone.
STANFORD, W E. 1910. 'Statement of Silayi, with reference to his life among the Bushmen'. *Transactions of the Royal Society of Southern Africa.* 1: 439.
STOW, G W. 1905 reprint 1964. *Native races of South Africa.* Cape Town: Struik.
STOW, G W & **BLEEK**, D J. 1930. *Rock paintings in South Africa.* London: Methuen.
SUMMERS, R. (ed) 1959. *Prehistoric rock art of the Federation of Rhodesia and Nyasaland.* London: Chatto & Windus.
THORNYCROFT, C. 1978. 'Mythical animals in the Marandellas district'. *Rhodesian Prehistory.* 16: 8-10.
TONGUE, M H. 1909. *Bushman paintings.* Oxford: Oxford University Press.
TREW, W I. 1984. 'Rock paintings at Sandhills in the Hex River valley, south-western Cape'. *South African Archaeological Bulletin.* 39: 130-37.
VINNICOMBE, P. 1960. 'A fishing scene from the Tsoelike River'. *South African Archaeological Bulletin.* 15: 15-19.
VINNICOMBE, P. 1976. *People of the Eland.* Pietermaritzburg: Natal University Press.
WILLCOX, A R. 1963. *The rock art of South Africa.* Johannesburg: Nelson.
WILLCOX, A R. 1956. *Rock paintings of the Drakensberg.* London: Max Parrish.
WOODHOUSE, H C. 1971. 'A remarkable kneeling posture of many figures in the rock art of southern Africa'. *South African Archaeological Bulletin.* 26: 128-31.
WOODHOUSE, H C. 1974. 'Creatures with both human and animal physical features depicted in the rock paintings of southern Africa'. *South African Journal of Science.* 70: 13-17.
WOODHOUSE, H C. 1975. 'Enigmatic line feature in the rock paintings of southern Africa'. *South African Journal of Science.* 71: 121-125.
WOODHOUSE, H C. 1978. *Rock Art.* Pride of South Africa series. Cape Town: Purnell.
WOODHOUSE, H C. 1979. *The Bushman Art of Southern Africa.* Cape Town: Purnell.
WOODHOUSE, H C. (Bert) 1984. *When Animals were People.* Johannesburg: Chris van Rensburg.
WOODHOUSE, H C. 1985. 'Elephants in the Rain'. *South African Archaeological Bulletin.* 40: 53-54.
WOODHOUSE, H C. 1986. *Comets in the rock art of southern Africa.* Collected papers in honour of James G Bain. Archaeological Society of New Mexico, Albuquerque.
WOODHOUSE, H C. 1987. 'Creatures with tusks in the rock paintings of southern Africa'. *South African Journal of Art History.* 2 (1 and 2): 40-46.
WOODHOUSE, H C. 1990. 'Utilisation of rock face features in the rock paintings of South Africa'. *South African Archaeological Bulletin.* 45: 112-116.
WOODHOUSE, H C & **LEE**, D N. 1976. 'The rediscovery of rock paintings recorded by George W Stow in the 1870s'. *South African Archaeological Bulletin.* 31: 29-30.
WOODHOUSE, H C & **LEE**, D N. 1978. 'Rock painting sites in the Mkhomazana valley, Natal Drakensberg'. *South African Archaeological Bulletin.* 33: 106-167.

INDEX

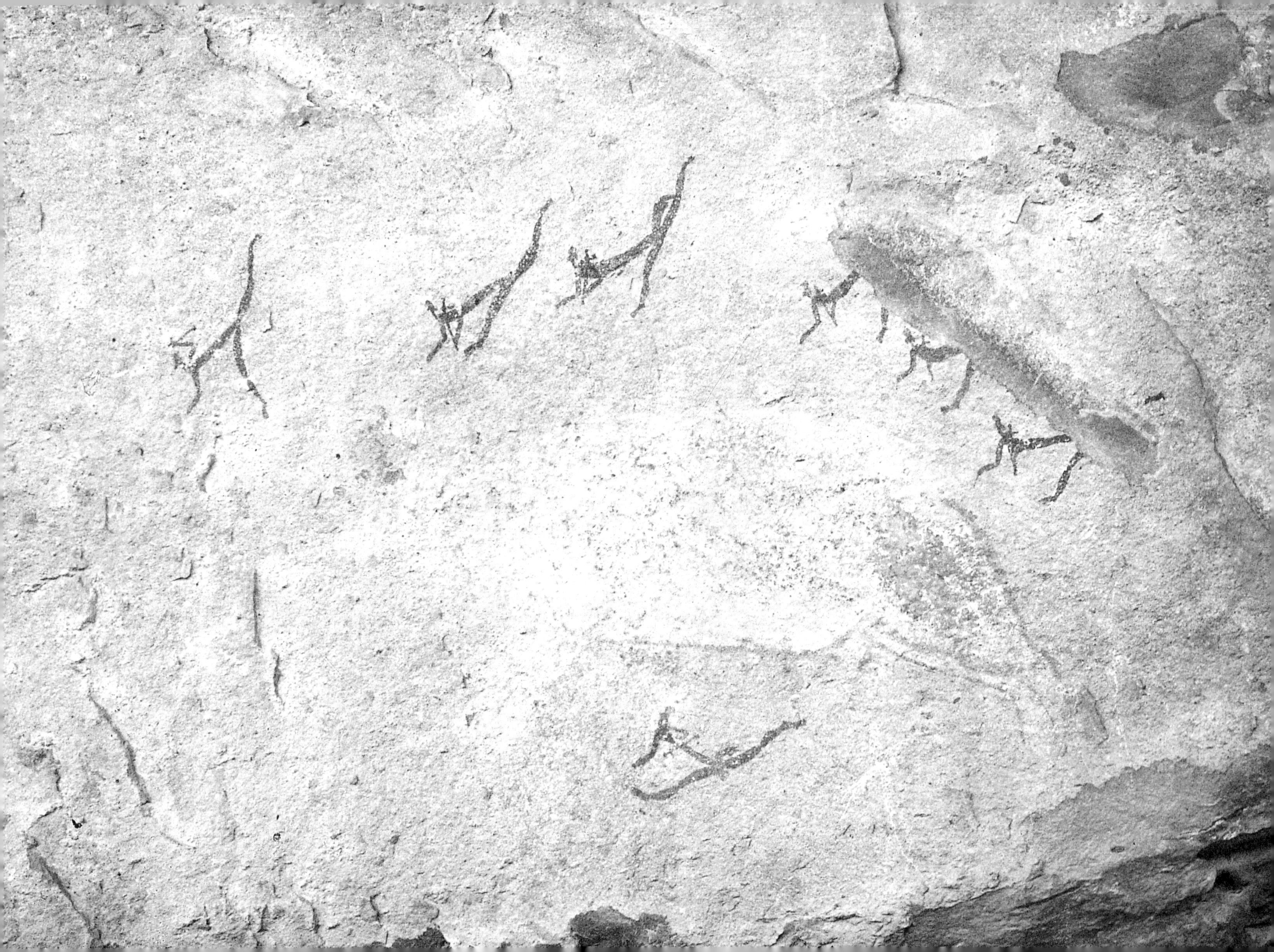